A Maths Teacher's 'Tricks of the Trade'

Rick Boyce

A Maths Teacher's 'Tricks of the Trade'

Originally a production of the Real Teaching Solutions Project Team

ISBN: 978-1-923533-22-6 (print)
978-1-923533-23-3 (ebook)

NATIONAL LIBRARY OF AUSTRALIA

A catalogue record for this book is available from the National Library of Australia

Armour

CONTENTS

THE TEACHING/LEARNING PROCESS IN THE CLASSROOM

TECHNOLOGY AND MATHEMATICS

GENERAL STRATEGIES IN MATHEMATICS TEACHING

ADDENDA

EXAMINATIONS and ASSESSMENT

WRITING ASSESSMENT TASKS

The process of learning to write and administer assessment tasks will always be a work in progress. There will always be ways to improve your assessment techniques. You can do this first by reviewing each item after you have completed marking your students' efforts. If you don't do the review and use the item/s again, you may perpetuate the mistake of using a poor assessment task.

Unfortunately, some teachers have believed in the past that their assessment task was perfect and that the students had not done the work to succeed. In fact, the opposite was often true. The assessment task had not shown what the students did know but what they did not know. This situation had often created a real fear of traditional examinations in the minds of students. The profession, in some cases, put it down to poor exam technique when the students, in reality, were given a poor assessment task. Therefore, it is imperative that we write assessment tasks that test what we teach and reflect how we teach.

For the young teacher starting out, remember your own experiences of assessment as a student to help guide you along the way. Don't just use items you have collected along the way or are given to use when you start your career. By all means, use what you are given as a guide. Alter it to fit into your teaching approaches. Have a teacher mentor on your school staff review it with you before you use it.

Always remember to make your first goal in all your assessment tasks to find out what the students do actually know. This will lead you to know what they need to know and your second goal. This will be to revise that knowledge to give your students what they need to succeed in the next teaching period. Lastly, always review each assessment task with the goal of improving it so you can use it again in the future.

What follows are suggestions of a general nature which will help you in an organisational sense to get your assessment development underway. There are no suggestions of a subject discipline nature. You will get these during your training, your teaching practice and from senior staff when you are appointed to your school.

When You Are In Training

Collect examples of as many different assessment tasks at all year levels you might teach.

Make sure you have an answer sheet and marking scheme/criteria to go with each item.

Write your own assessment tasks to test out the success of your practice teaching program.

Review each assessment task you use and rewrite it as necessary. Make notes on how to administer it as well as how successful it was.

Organise an assessment folder in year levels and subject disciplines.

When You Begin Teaching

Continue to gather assessment tasks as you go for your assessment folder.

Most schools and/or teachers will have past copies of all their assessment tasks on file. Sometimes, these will be on the staff or subject area school intranet. Consult and use these as initial guides to gain an idea of the standard of questions you might ask.

Don't just copy items or tasks or exams in full. Do this only if it fits what you want and reflects your teaching approach for that topic.

- # You can use old test items as practice tests for your students. These practice tests will provide ideas for your own tests if you mark the practice tests yourself.
- # Decide beforehand what skills/procedures you want to test. These should be the basic skills/procedures on which the topic is built.
- # Initially, write the assessment items to test these skills and procedures.
- # Now, write items that bring these skills and procedures together to complete more complex tasks. These could be tasks of a problem solving/critical thinking/ higher order thinking nature.
- # You must make sure that you have a range of questions from the simple to the difficult/complex. This gives every child the chance to succeed in some way.
- # Putting it a different way, your assessment items must test what basic skills and procedures they know and, later, in the assessment items, how they apply these together in more complex situations.
- # Do the assessment items yourself from the printed draft to test their validity and your suggested time requirements. Look for printing errors, errors of meaning and so on.
- # Amend the assessment items yourself. Then have another teacher proof read it and comment on it.
- # Do an answer sheet/model answer and determine the marking scheme or criteria. Put the marking scheme/criteria on the assessment tasks.
- # Write a clear set of instructions on page one of the printed paper. Include the time limit as well.
- # A rule of thumb that was suggested to me, early in my career, was to multiply the time you take to do an assessment task or examination by three to get an idea of the time students would take to complete that assessment task. Use this as the time you would give the students to complete the task.

ADMINISTERING A CLASS EXAMINATION

This chapter on class examinations concentrates on giving teachers the knowledge they need to administer an examination to their class successfully from the start to the finish. It looks at how the teacher goes about starting the examination; what he/she does as the students do the examination and how to wind up the whole process with as little stress as possible to all participants.

Starting the Examination

These are the steps to take as you prepare to begin the examination.

Line up students outside the room.

Give any instructions they need, for example, where to sit, how to ask questions.

Students are told what they are allowed to bring into the room.

Any belongings they must bring into the room are put on the floor at the front/or back of the room or on the floor under their chair.

Allow the class to enter the room under your instructions.

They write their name and yours on the writing paper/examination booklet and organise their gear ready to start.

On your instruction, they turn over the exam paper and with you check that it is correctly printed. Then they fill out their name and the class teacher's name on the front page using upper case/capital letters. (Their names are then easier for teachers to read).

Indicate any special instructions as well as the time for the test.

Start perusal time if available.

Start the exam reminding them of the time length and the actual finishing time.

Discourage borrowing of equipment. If it is necessary, the teacher should organise and supervise it.

Mark your class roll. Check this later against the actual exam papers you get as occasionally a student may deliberately withhold their exam paper but claim they handed it in and you must have lost it.

Remind the class that you will put a time line on the board and will mark off the time as the exam proceeds.

Don't allow students to have any other items on their desk except those items needed for the examination. (Be careful to make sure no one smuggles in writing paper that could contain answers and so on).

During the Examination

The teacher should:

Wander, at random, around the class to show you are actively supervising. If you need a rest, sit at the back of the room where you can see the whole class but they cannot see you.

Simply walk up behind them and watch if you suspect that a student is attempting to cheat. Ask them if they have a problem to give the impression you are aware of what they are trying to do.

Continue to mark off the time, regularly.

Warn the class that the time is up in 5 minutes.

If you have not already done so, collect any unused examination papers and writing paper/booklets.

Finishing the Examination

The teacher should:

Warn the class that time is almost up, that is, 5 or 10 minutes to go.

That 'stop work' means no more writing, just sorting out of the pages, stapling.

Remind the class to check their name is on every page.

Remind the class that no talking is allowed until all papers are collected or until they are dismissed and are outside the room.

Remind the class to ensure that the pages of their examination are placed in the correct numerical order before stapling.

Collect the exam papers row by row at the front door where you staple the pages together, if necessary.

Or collect row by row at the desks and staple as you go as an alternative.

Or have students leave them on their desks for you to collect at your leisure if the exam goes right to the end of the lesson time.

Count the number of papers you collect as you go to make sure all students hand in an exam paper even if they did nothing except write their name.

Collect all excess exam papers and writing paper. (This can be done during the exam).

After the Examination

The teacher should:

- Check that the names on the exam papers match with the names on your roll.
- Write the names of all absent students on spare exam papers ready for you to administer these exams to the absentees at your first opportunity or when specified by school policy. Note the date on the paper. When the paper is done, note that date also.
- Review how your organisation of the examination went and look for ways to become more efficient in its organisation.

You will find, as you become familiar with these steps, that your students are less stressed; the examination runs smoothly; no examination papers are lost and you walk out of the examination room feeling the glow of a job well done.

STUDENT BOOKWORK IN MATHEMATICS

It was school policy in my last school that students in Years 8–10 have a work pad for each subject. In the Senior School, a large folder was acceptable. In Mathematics, a work pad was considered to be more appropriate in the Senior School. Below are some suggestions I made to teachers in my Mathematics Department Handbook with regard to how students set out their bookwork for maximum benefit to them.

What is also important to note here was, at that time, in the senior Maths syllabuses, there was a communication criteria as part of the overall assessment procedure. Therefore, it was important, right from the outset of high school, that students were taught how to communicate their Mathematics effectively. The bookwork suggestions below were the beginnings of that process.

Setting-Out Suggestions

Although teachers may have their own ideas on setting-out, the following are some ideas to consider:

1. Pages might be ruled up in the following ways to facilitate logical setting-out.

   (a) Divide the page into 3 sections: a narrow margin on the left; a small working column on the right and a larger section in the middle to show the logical setting out.

(b) The second option is to divide the page into two halves to show the working in a logical way. Add a narrow margin on the left of each half. (This option would be useful in middle school classes where working out is less onerous than in more senior classes).

(c) In each case the student would write the date; textbook page reference; and the number of the exercise at the top of the page. The number of the question being attempted would be put in the margin. This helps students and teachers to keep track of their work for revision.

(d) Lastly, the student would rule a line across the page under each completed question. (Some teachers, often, asked their students to underline the answer as well.)

5. Encourage students to work down the page rather than across. This makes checking easier and reduces copying errors.

6. Neatness is a must with particular stress on numbers, symbols and algebraic terms to prevent errors in recognition that lead to wrong answers.

   Bookwork Assessment was part of the Assessment Program in the junior high school classes in my school. Consequently, there was two important points involved in this assessment.

   # The first was that all work had been checked and correct.

   # The second was that all working and explanations should be shown logically and clearly.

It is important that teachers check progress in this area in the junior classes. It could be on a formal basis, (that is, taking up the books of all the class to check) or on an informal basis, (that is, doing random selection of students' work).

Teachers were expected to date and to initial books as they saw them and offer written and verbal advice to the student on the student's progress.

A final comment:

Below, for your consideration, is a notice on bookwork I saw outside a junior class aimed at encouraging students to treat their bookwork seriously.

> *Bookwork must always be presented neatly.*
> *Take your time to write neatly and set out your work correctly.*
> *Be proud of the presentation of your written work.*

ASSESSING STUDENTS' BOOKWORK

Some school assessment programs in middle school and junior high school include a mark for bookwork as part of the reporting process in some, if not all, subjects studied.

Below are a set of criteria that I have used to assess student's bookwork in junior high school Mathematics.

The criteria are:

1. Headings, references and dates are stated for each lesson. For example:

   Monday Classwork/Homework

   5/6/13, Future Maths p.238, Ex52, No.7–9

2. Working and explanations are clearly and logically shown.
3. Work has been checked and corrected if mistakes are found.
4. Cover of the workbook is neat and labelled with

   Name, Form and Teacher.

5. Hand-outs are glued into book in the correct place.
6. Book is correctly ruled up and work is neat and legible.

At this point, I would like to stress the importance of Criterion 1. This is particularly important for students who miss class time through illness. By checking the student's workbook against another student's workbook the teacher and the sick student are able to record what was missed during the student's absence and organize a 'catch-up' program.

Mathematically speaking, Criteria 2 and 3 are the most important. Some teachers may give greater weighting to these two criteria in assessing bookwork.

The teacher can also train the students to do a self-evaluation of his or her bookwork or have a peer do it for them. Below are the items in a self/peer evaluation to check off with a *'yes/no'*.

Self/Peer Evaluation

1.	All work checked and corrected.	*yes/no*
2.	Working and explanations shown.	*yes/no*
3.	Ruled up, neat and legible.	*yes/no*
4.	Headings, references and dates included.	*yes/no*
5.	Hand-outs glued into book.	*yes/no*
6.	Book cover is neat and labelled.	*yes/no*

Standard

Almost all ***YES***	Very Satisfactory
YES sometimes	Satisfactory
Mostly ***NO***	Unsatisfactory

With my classes, these self/peer evaluations were done a few weeks before my formal evaluation of the student's bookwork to encourage the students to improve their bookwork. The evaluation criteria that were unsatisfactory indicated to the student where improvement must be made.

TEACHING STUDENTS TO COPE WITH STATE/NATIONAL TESTING PROGRAMS IN MATHEMATICS

Throughout the education world, students are subjected to testing programs at the end of their school years that decide whether or not they gain university entry. In some jurisdictions, these tests are marked using a criteria-based system. Often these test what are called 'core skills' as opposed to the content, basic skills and problem-solving inherent in the traditional school assessment program. (Often, the depth of material is based on the standard expected of students a couple of years below the graduating class).

Students often find the different approach to assessment in these 'tests' difficult to adjust to. Consequently, schools give students practice in answering questions tested using criteria.

What appears below is the advice I would give my senior classes leading up to the practice tests and the 'real thing'. Even though it is Maths in theme most of the advice expressed here would help students in all subjects tested in these sorts of state/national testing programs.

These are the ideas that I stress to my students:

- # Everything you need to solve the problem is there in the question or associated material.
- # Simple ideas are often the starting point. **Don't** overlook the simple. Look for the obvious first.
- # Begin with the end in mind. Work backwards. (Think about how you solved geometry problems by looking at what you had to prove last).
- # Underline the key words that tell you the 'core skill' being tested, for example. calculate, evaluate...
- # Treat all questions as if they were easy.
- # Don't linger at a difficult question in a unit. Simply, go on to the next unit. Remember each question tests a new core skill that may not be related to the previous question.
- # Use the tables and diagrams provided to help you. Draw new ones if that helps.
- # Try every question.
- # Make sure your answer is realistic. Remember Maths is practical and, more often than not, life-related.
- # Use your normal Maths checking process as you go. Do it step by step to save time.
- # Remember to do only what the question asks you to do—no more, no less and not what you would be expected to do in an ordinary Maths exam.
- # Remember, too, the Maths involved is at a standard a year or two below your level so that should not prevent you attempting to answer the question.
- # Finally, even though the context of the question might not seem to be Mathematical, remember it is a test of the core skills associated with Mathematics.

After the practice test has been marked and returned, I would always review with my class how well they put the ideas expressed above into practice and how they could improve their use in the 'real thing'.

CRITERIA-BASED MARKING IN MATHEMATICS

Traditionally, assessment in Mathematics at high school and beyond has been the timed, formal, pen-on-paper examination. Often, talented Maths students have failed to show their expertise in these formal examinations due to exam nerves. Other students who learn in a non-maths/logic way found that the formal situation did not fit into their learning styles. What has further pointed Mathematics teachers to use a variety of pedagogies and then non-traditional assessment techniques has been the recognition of the existence of Multiple Intelligences. This suggests strongly that students learn in different ways. Additionally, more students are staying at school longer and studying Mathematics in the hope of gaining entry to more prestigious university courses.

These factors have lead in some educational jurisdictions to new Mathematics syllabuses that require assessment tasks to reflect a wide variety of pedagogies as well as the use of computer and calculator technology. The traditional formal examination remains—but has become one of many assessment techniques that may be used. This has led to criterion-based marking in the non-traditional assessment tasks.

This chapter looks at one alternative assessment item that uses criterion-based marking. It is summarised and the criteria are included and some further explanation added.

Year 12 Maths B Assignment
NETWORKS — CRITICAL PATH ANALYSIS

Select one of the following projects and use Critical Path Analysis techniques to complete the questions below:

A ***Compiling School Magazine***

B ***Organising School Musical OR Rock Eisteddfod***

C ***Organising School Senior Formal***

What is above was the basis of the assignment. What followed this is the general description of what was expected in this assignment. Then the questions to be answered with regard to the Critical Path Analysis were specified. They included questions which tested their basic understanding *(Techniques)* of Critical Path Analysis plus four questions of a problem-solving *(Applications)* variety. Two were regarded as simple and two as complex questions.

Here is an example of one of each of these.

A *techniques* requirement was:

Construct a network for these activities.

An *applications* (problem-solving) question was:

Select a completion date for your project and show justification mathematically for the latest date on which you could begin the project.

The assignment was marked using three criteria. They were:

1. *Communication* (This criterion looks at how the student communicates Mathematical working).
2. *Techniques* (This criterion assess the student's knowledge of the core ideas/ learning work of the topic) and

3. *Applications* (This criterion assesses how the student applies knowledge of Mathematics to a problem-solving situation, often in an unfamiliar context).

Obviously, the syllabus documents will specify and define the assessment criteria applicable in each system and how they are applied. I simply use these as an example.

Below are the assessment criteria I used for this assignment.

MARKING CRITERIA

COMMUNICATION:	***very good*** ***satisfactory*** ***unsatisfactory***
PRESENTATION:	***Sentence construction*** ***Spelling*** ***Use of Mathematical symbols***

Allocation of Standard

H Mostly very good

S Mostly satisfactory

U Mostly unsatisfactory

TECHNIQUES

The student has correctly:

(a) Recorded activities;

(b) Recorded time required to complete each activity;

(c) Ordered the activities—stated in writing which it precedes;

(d) Drawn a Network/s as appropriate;

(e) Drawn Precedence Boxes;

(f) Completed a forward Scan;

(g) Completed a backward Scan;

(h) Drawn in the Critical path;

(i) Drawn a Gantt Chart;

(j) Significant use of the Internet including listing of URLs.

Allocation of Standard

A Completed ALL the above activities correctly

B Completed (a) – (h) correctly

C Completed (a) – (h) maybe some minor errors

D Some attempt

E No attempt

APPLICATIONS

F No attempt

E Little progress

D Substantial progress but no solution

C Incorrect solution caused by a major conceptual error

B Correct solution and calculations shown (minor errors allowed)

A Correct solution and justification of techniques used.

Notes:

In Techniques and Communication, you will notice all the basic skills are listed. This enables the student to know the least requirements to gain a satisfactory grading.

In Applications, the criteria are generic so as to not lead the student towards a solution. This makes it a real test of how the student uses his/her Mathematics knowledge. Specific marking criteria are developed for the marker to use and these are published for the students after the assessment task is returned to them.

In Applications, you will notice the words 'simple and complex' to describe the problems. The syllabus required a variety of unfamiliar problems from simple to complex in Applications to help delineate between students at the top end of the scale.

One of the important advantages of this type of criteria-based marking is that you can use the same criteria over and over once you have revised them. All you do is to change the question.

Lastly, it is important to note that in discussing the assignment initially, the teacher should remind students to address the criteria at all times. This will ensure that they get the maximum grading for each criterion that they are capable of getting.

STRATEGIES FOR THE IMPLEMENTATION OF NON-TRADITIONAL/ALTERNATIVE ASSESSMENT TASKS

Let me begin by defining Alternative/Non-traditional Assessment as any assessment task that is not the traditional pen on paper examination or a written assignment in a non-practical subject.

The last 14 years of my permanent teaching career were spent introducing and implementing an Alternative Assessment Program throughout Years 8 to 12 in my role as Head of Mathematics. This was a major shift in the way Mathematics' teachers assessed their students.

Their assessment had traditionally been pen on paper examinations of a fixed time. This change came about because the new syllabuses that were introduced required a variety of assessment techniques to be incorporated in the school's assessment program for the school to have its program accredited.

Another factor in the 'equation' was the arrival of technology in the form of graphics calculators, computer software and internet which all lead to a revolution in what could be taught and how it could be taught in Mathematics.

In addition, most students were remaining at school longer and studying Mathematics. This meant that Maths teachers needed to realise that their students were not all Maths-logical thinkers and thus required a variety of teaching approaches to cater for their different learning styles. The syllabuses recognised this and encouraged teachers to use a greater variety of teaching approaches. This meant that teachers needed to use a variety of assessment techniques to reflect those teaching approaches.

The introduction of other assessment techniques other than the traditional pen on paper examination is a long-term development project for teachers and must begin with small steps. Therefore, I began in a year level where the reporting pressure was the least. This way the results of a poor assessment task could be discarded without any angst. The lowest year level was the ideal place to start allowing you to progress up the school getting all the staff involved over time.

I took my school immediately into the new assessment regime when it was to begin to allow access to the available specialists to act as our guides. Later, as a result, we were asked to help teachers in other schools with their implementation of the new assessment techniques.

Below is a summary of advice we gave these teachers.

Curriculum Heads need to be the innovators. They must write and use the Alternative Assessment tasks with their own classes first. (That means, of course, they need to be involved in the teaching at that level). Then, with instructions and advice on how to use them, the task is given to other teachers to help them get started. In our first year I did that.

In my department, **each year level or subject had a coordinator.** Their responsibility was to plan the year's work and write the assessment tasks. Therefore, in year two of our development, I organised the Year 9 program and a senior staff member took over the reins in Year 8. Since most Maths teachers taught in Year 8 or 9, each teacher was assigned an assessment task to write. This was to get everyone involved in the process as soon as possible. We continued that process until all year levels and subjects were incorporating Alternative Assessment tasks.

Make sure you select units of work that lend themselves to assessment approaches other than the traditional. There will still be sections of the course where the traditional pen on paper examination will need to be used. These traditional examinations will always be part of a valid assessment program.

Once the ice is broken, **all staff must be involved in writing, using and marking these types of instrument (ASAP).**

It is important that **each instrument be evaluated after it is used;** re-written with advice for future use. Don't discard an assessment task if it does not work well the first time. Remember we are all novices at writing and using Alternative Assessment techniques.

Most Alternative Assessment Instruments take longer to do than expected, especially if the teaching approach and the instrument are not 'in tune'. **In your early test instruments, restrict the amount you test.**

Marking is often protracted. The best way to do it is to mark one section of all the students' assessment tasks. This ensures that you are consistent with the way you mark all the students' work. It also allows you to find alternative approaches that are correct or acceptable. Add these other approaches to your marking scheme/criteria.

Include general marking criteria for problem solving tasks. Specific criteria published with the instrument often lead the students to a solution/approach. This reduces the problem component of the task.

Students must be given practice in Alternative Assessment techniques to ensure they are not disadvantaged by the process.

It is important for the Curriculum Head to be careful in their selection of who are the first coordinators they appoint to help start the process. They must be committed or the implementation will stall.

Make sure all staff members are involved in any discussions about the assessment and that there is flexibility in how the implementation proceeds.

Finally, **provide your staff with as many professional development opportunities as you can** to give them the knowledge to proceed.

THE SECRETS TO SETTING SUCCESSFUL PRACTICE EXAMINATIONS

Students often don't treat these practice examinations seriously because they are not for 'marks'. However they are an essential part of exam preparation. It is important that they be given while you and the students have time to address deficiencies in their learning. They must be factored in to your learning program.

The following ideas are important to consider:

1. The test should cover each area studied and to be tested.
2. Questions/tasks must be organised from simple to difficult starting with skill or learning work examples before you add problem-solving/higher order thinking examples.
3. The standard of this practice test must reflect the standard of the actual test itself.
4. Organise it as an exam situation, that is:
   (a) Spread out the desks to prevent cheating, talking and so on.
   (b) Encourage students to follow the exam technique that has been taught.
   (c) Tell them to try every task and continue to try until the end of the test time.
   (d) The practice seating must be in exam mode.
   (e) Write time intervals of ten minutes on the board and cross out the time elapsed to mirror the time stress of a real examination.

5. Make sure there are enough tasks to cover the time available, that is, add time pressure.
6. Teach students how to do an exam the best possible way to gain the best outcome. (You would do this as part of your examination technique training).
7. Seek out past papers and use them as part of the practice test. This gives your students an idea of the type and standard of questions to be asked. You won't need to make up too many questions but make sure none are actually on the real test. Also, answers will already be available in your examination file (if you have one).
8. Cut and paste the answers and publish them as a model when you have reviewed the practice test. You could give each student a hard copy; publish the answers on your notice board or perhaps publish them on your home page on the school intranet.
9. When you review the practice test:
   (a) Begin by giving the answers or a brief explanation of how each task should be done.
   (b) Then survey the class to see which tasks need to be fully worked or explained and reteach where necessary.
   (c) Point out common and/or expected errors which might include carelessness, omissions, not knowing rules, poor reading of the task and not answering the question fully or correctly.
   (d) Have students write down the types of errors made and determine how many marks they have lost. Discuss how to avoid these and make the students aware of their possible result had it not been for these errors. Often poor exam technique is the main cause.
10. Develop in your students a belief:
    (a) The exam is not over until the time is up.
    (b) Try and try and retry all unfinished tasks.
    (c) List any rules or procedures you know and try them.

(d) Write down the data from the task as a starting point.

(e) Underline key words (often verbs).

(f) Circle what you need to find or do.

It is important to remember that the practice examinations help you and your students find their deficiencies but they also help develop examination techniques and help reduce the stress of the real examination. This can only lead to results that better reflect a student's learning and developed skills.

THE SECRETS TO SETTING SUCCESSFUL PARALLEL EXAMINATIONS

Often, especially in secondary schools, there are subjects such as Maths and Science where not all classes are programmed at the same time. This means period tests cannot be done at the same time either, creating a security issue. It is one often complained about by parents who claim some students had an unfair advantage by supposedly knowing in advance what was on the assessment task. This raises the issue of setting parallel tests and this is not as easy as it looks.

Hints for Writing Parallel Tests

Write the original test as your starting point.

- # In subjects like Maths, Accounting and in some Science questions, you can write the second test by simply changing the data in each question so that the skill or understanding you are testing remains the same only the working and answer changes. This establishes the validity and fairness of the parallel test.
- # Another way is to use your word processor to your advantage. Simply rearrange the order of questions as well as page changes to give the impression that they are different. Call them test A and B. This can be done in all subjects.

In subjects where diagrams/pictures are used, put them on a different side of the page. You might even be able to rotate or reflect the diagrams without altering their impact on the question.

For different class groups, use different coloured paper with the same test as a ploy. Still, call them test A and B. You can get away with this ploy in middle school/junior high school.

Even add a picture or two to change the whole look of the exam paper or change the ones you are using. You could make them larger or smaller or put them in a different position.

In some subjects or test instruments, you are testing a particular skill or set of skills. Here you are using a particular scenario and a set of criteria to assess the student. Therefore, the question or procedure does not change only the data used. Here is an example of what I mean using a Maths context.

In a statistical investigation, students are given a set of data on aboriginal artifacts (axe, heads, spear ends, dillies) found in a disputed area. Four tribes claim ownership of this land. Data on similar artifacts from each tribe are given to the students.

Their task is to use statistical techniques to investigate the data of each claim and using their statistical calculations to put a case forward suggesting which tribe statistically has the best case to be the traditional owners. You could give different data to different classes.

Each student will approach the problem differently because of their own experience and level of understanding. You could even allow students to sit side by side with different data but the same question.

There will still be complaints. Simply explain to the students (and parents too, when necessary) that the test will be of equal standard. Explain also that where cut-offs are used, they will be based on each individual test, for example, 80% might equal an A. If anomalies are found in the marking of the tests, adjustments will be made to ensure comparability. Parents can be given more detailed explanations, if requested. It is always politic to inform your principal or team leader as well as parents when parallel tests need to be used.

It is possible to use criteria to assess test instruments instead of marks when you use parallel test instruments. Because you are assessing the level of skill using the criteria, then a perceived difference in test difficulty will not prevent the valid assessment of the student's work.

Obviously, you are only going to use parallel tests where all students cannot sit for the assessment task at the same time. If you feel that there may have been a 'security breach' and some student/s have performed above expectations, don't discard the assessment task outright. Simply give the suspect student/s a short test on the same topic with significantly different questions to check that their marks on the original task are valid.

A GUIDE TO REVIEWING ASSESSMENT TASKS

Returning assessment tasks and examination papers to students to check the marking and to review their mistakes is an essential but taxing task. It requires careful planning to make it an efficient procedure from which students can learn from their mistakes. The process, essentially, is the same in all subject disciplines whether they are high school subjects or the primary classroom.

Often, students have a negative approach to such reviews. To counteract these negative feelings I give the students a number of reasons why we should have them.

The reasons are:

1. I want to make sure the students get the marks they deserve. They do this by:
   (a) checking that all questions have been marked as well as marked correctly
   (b) checking that my addition of their marks is correct as well as the allocation of an appropriate level of achievement/grading.

3. Students have an opportunity to learn about mistakes they have made which can be eliminated in the future. (They also find out how many extra marks they might have gained without those mistakes.)

4. Students have an opportunity to ask questions about the marking scheme/criteria.

5. They have an opportunity to find out which basic skills they need to relearn for the future.
6. They will also learn how well they employed the exam technique I have taught them with a view to further improving that technique for the next formal assessment instrument.

My Pre-Planning

Below is a list of what I do in preparation for these reviews.

Write a list of questions/tasks I need to review, that is, discuss or work out fully for the class.

Write a list of errors of carelessness, procedures or learning work that I need to highlight.

Write appropriate comments/advice on each student's assessment tasks aimed at how they might improve. I point out careless errors, errors of omission and learning errors. I praise good work.

Record all results in my mark book initially in pencil until I confirm their results in the review process.

Make a list of areas that need to be retaught.

My In-Class Review Strategy

To overcome the negative feelings of these reviews, I have developed, over many years, a procedure that is efficient, uses less time and allows students to get the marks they deserve. The process also validates the reasons for doing the reviews.

Below I detail how I go about these reviews.

Step 1

I give the following instructions to the class before I return their assessment. These instructions are designed to make sure the students get the correct marks, the assessment tasks are reviewed efficiently and the students have an opportunity to learn from the review.

1. Check that each part of each task has been marked.
2. Check that the marks tallied for each question are correct and that the total mark for the whole task/exam is correct.
3. If there is a problem, the student is to put a question mark in pencil near the 'error' in the margin and one on the front page of the task/exam so I know it needs to be reviewed. I will personally check each of these later.
4. Under no circumstances are students to check, compare or look at anyone else's work. They are to concentrate on their own work. I tell them if they miss finding my mistake/s during the review then they will get no credit for it if it is discovered later. (I am, of course, kidding if it is a formal assessment item).
5. They are not to change anything on their papers during the review. Any changes or additions will result in zero marks for that part. The only marks they can make are '?' marks.

Step 2

1. I give the class the answers to each question or discuss how to answer it. The class checks their answers as we go. I do not accept questions at this point. I suggest they write down any questions to ask me later.
2. I review the marking scheme for each question/task indicating how marks were allotted. If there are marking criteria, I explain how they are applied to each question/task.
3. Once the above steps are completed, I ask for questions. I will explain or fully work any questions/tasks as required.

Step 3

1. I collect all the assessment tasks getting those ones where students believe I have made an error, first.
2. I re-mark all the assessment tasks that have been queried.
3. Then I finalise all marks or ratings in my mark book or computer file.
4. At my next lesson, I confirm with students the marks/rating they received.
5. Where students' scripts must be kept for reporting purposes, I file them away immediately.

A GUIDE TO TEACHING STUDENTS HOW TO GAIN GREATER SUCCESS IN FORMAL EXAMINATIONS

Introduction

I grew up in the years when the scholarship examination at the end of primary school was still an important step in the education process. Each Friday morning, the scholarship classes—the final year in primary school—had a two-hour test examination on English, Mathematics or Social Studies. At high school, there were many 2 and 3-hour exams at the end of each term leading to the formal external Junior and Senior examinations. Most of my first degree assessments at University of Queensland were 3-hour exams. As an evening student, it was essential to develop good exam technique. This article reflects these experiences.

Preamble

Up to 50% (sometimes more) of the test instruments used in schools are the traditional pen on paper examination held under strict examination conditions. For many students this can be a stressful situation where they do not perform at their peak. Apart from a good study program, the best preparation is to teach the student how to best do the actual examination or test.

Below are the steps I teach my students from Term 1, Year 8. I repeat it prior to every examination right through to the end of Year 12. This procedure can be used in all year levels and subjects successfully. In primary school, teachers might introduce these ideas slowly and actually go through a practice exam explaining to the class how to do these steps. I have even done this with my secondary classes. Remember it is an on-going process.

Firstly, I remind them of these issues:

1. Get a good night's sleep.
2. Have a good breakfast/lunch, etc.
3. Drink plenty of water.
4. Don't study 'madly' during the last 30 minutes before the test.

All these ideas are to ensure the student is in the best physical condition possible to do the exam.

Examination Technique

STEP 1: Read the examination/test paper at least twice.

(a) Decide which questions are easy and more difficult. Mark them difficult or easy.

(b) Decide on the order that you will do them.

(c) Do the easy questions FIRST.

(d) Why? They take less time; therefore you gain extra time for the more difficult questions.

Successfully doing a question early creates confidence.

You don't miss out on marks you can get.

You give the impression to your marker you know your subject.

Write out a short plan of how you are going to answer the question.

STEP 2: *Work out a time/mark allocation.*

(a) This gives you a rough estimate of the time required per question. Use the first few minutes to plan; the last few to check over and edit where necessary.

(b) Allow time at the beginning of the exam for reading and at the end for checking.

(c) If you are doing a more difficult question; Don't spend all of your time on it if you are not succeeding.

(d) Go on to others. When you return; read the question afresh and what you have done before you continue.

(e) Continue to try all unfinished questions to the end of the exam/test time.

STEP 3: *Checking is a compulsory part of every test.*

There are several ways to check:

(a) Check you have copied down all the correct data for the question.

(b) Check that your answer (its size, etc.) fits in a practical sense into the scenario of the question.

(c) Check, in fact, that you have actually answered the question fully.

(d) Towards the end of the exam time, check you have done or tried every question and every part of each question.

(e) In Maths questions or questions involving calculations, check every step as you do it.

STEP 4: *Make sure you have been:*

Neat

Tidy

Organised

Logical

Clear

Concise

AND you have

correctly numbered questions

your name and teacher's name appear on each page.

In other words, make sure it is easy for you and your examiner/teacher to understand/ follow what you have done. This gives the examiner/teacher every opportunity to give you the marks your efforts deserve. It also gives you the best opportunity to check your answers and ensure they are correct and that you get the most marks possible.

Remember:

Have with you biros, pencils, eraser and ruler.
(Coloured pencils/biros can be useful, too.)

Don't forget the calculator. Make sure it has charged batteries.

HOW TO ANSWER QUESTIONS IN A FORMAL EXAMINATION

Advice for the Student

This is a generic approach to this topic.
The next chapter extends it into a Maths context.

There are many able students who do not perform well in traditional pen on paper formal examinations. It many cases, it is simply nervousness. By having a plan of attack that a nervous student can adhere to, results can be improved considerably.

The student needs to practice this plan of attack in every lead-up examination they do to inculcate the practice in their psyche.

Below are a set of strategies for you as a student to adopt when doing formal examination questions.

1. Read the question once to get a general idea of what you are asked to do or solve.
2. Read the question again underlining or highlighting the 'doing' or 'action' words because they tell you what to do, e.g. solve, compare and contrast, write an essay, list.
3. Write down all the data you are given to answer the question.

4. Write down any rules, axioms, or procedures you might use.
5. List or plan the steps you might use to answer the question or solve the problem.
6. Write down what you need to find/do to provide the answer, i.e. the end point.
7. Now, do the question.
8. Use a checking process to ensure you have done what you have been asked to do.

Other hints to consider:

1. In some subjects like Maths and Accounting, everything that you need to solve the problem is in the question itself.
2. Simple ideas are often a good or the best starting point. Don't overlook the simple. Look for the obvious.
3. Begin with the end in mind. Work backwards from the answer.
4. Treat all questions as if they are easy. Don't expect them to be difficult.
5. Even if it is difficult, still follow the steps above that you have practised and try them.
6. To get a start in non-Maths/Science subjects, use 'how, when, where, why,' OR 'past, present, future,' OR 'local, state, national, international,' OR 'before, now, after,' as catalysts to get going.
7. Diagrams, tables, timelines and lists can help. Create your own and use them to help answer the question.
8. Make sure your answer is realistic, i.e. it is a real-life solution.
9. Don't do more than you are asked to do as you get no more marks and you use up valuable time needed elsewhere.
10. With a checking process, it is best to have a step by step one as this saves time often preventing unnecessary extra work.

Finally, all these hints and strategies will help improve the writing of each examination question.

To ensure that you get the best result possible, there is one other piece of advice you need to follow. You must decide during your perusal time, which questions are the easiest, that is. you know most about, and which are hardest for you.

Answer the questions doing the easiest first and the hardest last. This ensures you will get all the marks you are capable of getting. You will start confidently. You will gain time as the easiest questions often take less time to do well. This extra time will help you give a better answer on the harder questions.

STRATEGIES FOR 'ATTACKING' MATHS PROBLEMS

Advice for the Student

In the last chapter, *How to Answer Questions in a Formal Examination—Advice for the Student*, I discussed how to answer questions to gain the best possible results in a test. This chapter continues that theme but this time in relation to answering questions in a formal Mathematics examination. The strategies mentioned in the previous article should be applied to the Mathematics examination as well as the ones discussed below.

It is important to define what I mean by a problem in Mathematics before you start to study the strategies to solve them.

These problems are almost entirely 'word' problems. More often than not, the student needs to use a variety of Mathematical skills or ideas to gain a solution. Often, particularly in the senior years of high school, there will be an unfamiliar context in which to use your Mathematical knowledge. Alternatively, there may be a series of sometimes complex steps necessary to achieve a result. Finally, the answer is not one which is obvious.

Below are a list of strategies that, if used together, will help you gain greater success in solving real problems in Mathematics not just ones you have practised. However, remember, if you don't know your basics in Mathematics then no set of strategies will help you solve the problems.

Strategy 1:

…is and will always be:

Know all your learning work and procedures as well as you can.

The remaining strategies are as follows.

Strategy 2:

Remember, everything that you need to solve the problem is in the question itself. (So list what data the problem gives you as your starting point).

Strategy 3:

Checking is a compulsory part of every problem you are to solve. Here is a checking procedure to use:

It is best to check as you do each step in the problem as this saves time often preventing unnecessary extra work.

Ensure you have done only what you have been asked to do. Check, in fact, that you have actually answered the question fully.

Check you have copied down all the data for the question correctly.

Check that your answer (its size, etc.) fits, in a practical sense, into the scenario/context of the question.

Strategy 4:

Make sure you have been neat, tidy, organized, logical, clear, and concise. This will help you with your checking and allow the examiner/teacher to follow your logic easily.

Strategy 5:

This strategy was mentioned in the first article. It is part and parcel of answering any examination problem, especially in Maths.

It is: *List the steps you need to take, in order, to gain a solution.*

Below is an example of what I mean by this strategy in Mathematics.

The Swimming Pool Problem

'How long does it take to fill a swimming pool with a bucket?'

Here is how I teach this strategy:

Step 1:

I write the above problem on the board.

When I do this, I ask the students for their reaction:

It will be: 'We can't do it.'

You ask: 'Why?'

Their reply: 'There are no dimensions.'

Your reply: 'You don't need them. If I gave you them to you, what would you do?'

Step 2:

Now I have the students write down the steps they would use.

Step 3:

Then I discuss the steps the class select and list the steps on the board.

Find the volume of the pool.
Find the volume of the bucket.
How long does it take to fill the bucket and pour into the pool?
How many buckets of water do I need to fill the pool?
Find total time to fill the pool.

Step 4:

Now, I make the point that the above steps do not mention the dimensions of the pool. It doesn't matter what it measures, you still follow the same process.

Step 5:

Lastly, I emphasise that a correct answer depends upon the correct steps, i.e. method of solution.

A student can't learn these strategies overnight and expect that they will 'come to you' easily in a formal examination situation. You must practice using them. Make a list of the strategies and have them with you as you try each new problem. Evaluate how well you use them and work to improve those you find hardest to use or are easily forgotten. Look to your teacher for help with this process. Remember, in an examination, be disciplined, write out the list of strategies you will use before you start and use them to solve the problems.

TEACHING STUDENTS AN EFFECTIVE CHECKING TECHNIQUE IN MATHEMATICS

Why is a checking procedure necessary in Maths?

Here are four reasons you can give your students.

1. It ensures that each student can achieve their best possible result.
2. It saves time in examination situations allowing the student to spend more time on other questions that can be attempted. This means the possibility of more marks and better results.
3. It helps maintain confidence.
4. It means that the students should never get it wrong if they know and understand the Maths involved. Subbing your 'answers' into each equation to make sure they make all the equations true is an example of what I mean here.

How do you go about developing a checking mentality among your students?

Below are the strategies I teach my students to develop an effective checking process in Mathematics. Many of these strategies could be effective in other subject areas.

Strategy 1: Preparation for checking

By this I mean that the teacher must inculcate a checking strategy as an automatic procedure within every student's consciousness. That means it is done with every exercise every day and not just done at assessment time.

Strategy 2: Setting out

Logical, clear and neat setting out makes it easy for students to check. Teachers should constantly model appropriate setting out for each new procedure.

Strategy 3: Working down the page

Logical setting out that goes done the page, line by line, makes it easy for the eye to check that the student has transposed the correct numbers or symbols.

Strategy 4: Step by step checking

This, I think, is the most important strategy in the process. The student must develop a mindset that does the checking line by line as each step/line is completed. This strategy saves time, maintains confidence and increases the success the student has in assessment tasks.

Strategy 5: Answering the question

(a) Did you do what the question asked you to do?

(b) Did you answer every part of the question?

Strategy 6: Estimating the answer

When the student reads the question, the type of answer required should become apparent. There should be an expectation of what is the expected answer. This allows students to compare what they obtain for an answer with what they expect the answer to be. This also allows students to consider the correctness of the answer that they obtained.

Strategy 7: Calculators and checking.

The student must remember that the calculator is only as correct as the data inputted and the keys punched. Therefore, every calculator use must be checked by doing the calculation again. Students must record the first calculator answer before they do the calculation check. If the answers are different, then a third calculation is required to find the correct answer.

Strategy 8: Copying data from the written question.

This is the first and often most common of all errors made by students. It is essential that copying down of the data is checked automatically.

Strategy 9: Diagrams.

Diagrams are an essential part of some Mathematical exercises. Transposing the data from the question onto the diagram needs again to be correct before a student can begin to solve a problem. Diagrams need to be large to be useful.

Strategy 10: Teaching where errors occur.

Each time a new topic is taught, explain to your students where common mistakes occur when doing exercises in that particular topic.

Strategy 11: The Assessment Review

It is important to make reference to where individual students make mistakes in their assessment work. It is important to point out, after each assessment task, where the errors were made and what checking strategy was required to eliminate those errors. Point out, also, how many marks that those avoidable mistakes cost the student. This should be motivation enough for students to adopt a daily checking regime.

Obviously, a successful checking process will not enhance a student's Mathematical ability but will always lead the student to the best possible results.

The TEACHING/ LEARNING PROCESS —in the— CLASSROOM

TEACHING MATHEMATICS IN THE 21ST CENTURY

Mathematics and learning your native language are seen as the cornerstones of education in our schools. They are the starting points.

However, for many students, Maths becomes a burden and something to be avoided, if possible. Often this occurs because of the way we teach Maths. Added to this is the fact that students are staying longer in school and Maths continues to be part of their curriculum. Many students would rather 'drop' Maths in their later school years.

This article advises on how we can make Maths more appealing to our students.

Below are thirteen strategies to use to help students want to be totally involved in their Maths development.

1. **Mathematics should be fun, relevant, life-related.** Use such strategies as a fun quiz, real-life questions, easy to difficult challenges, unfamiliar contexts and speed tests to name just a few.
2. **Teach Mathematics the way you would like to have been taught,** not as you were taught, that is, remember that you were often bored and you could not see the relevance of Maths to your life. Don't allow your class to feel that way.
3. **Use a variety of teaching strategies that fit the topics you are teaching.** (Don't let Maths be just chalk and talk and practice of multiple exercises). Assess each topic in a way that reflects your teaching approach. Use technology, cooperative learning techniques, hands-on material, practical lessons, the quiz and any strategies that take into account the different learning styles of your students.

4. **Teach Maths through Stealth.** The quiz is a way to create learning by stealth. It seems to be more like fun than learning Maths for many students.

5. **Use your students as assistant teachers.** More able students are better at some aspects of Maths than others. Use them as mentors in their areas of expertise. You may need to give them some training but you will find that the students react well to their help and progress faster. What is important about the mentor's words is that it is in the language of the student. This enables the less able student to understand more quickly.

6. **Teaching Maths should be challenging, exciting and fun to you, the teacher.** Look for real life examples to use in your teaching and assessment. Include a short problem-solving/critical thinking exercise in every lesson. It does not need to be difficult every time; for difficult ones, give clues slowly.

7. **Experiment, evaluate, review, plan and try again.** Introduce new teaching strategies into your program and perfect them with a review process. These different strategies will better cater for your students' different learning styles as well as adding new, interesting, teaching challenges for you, as the teacher.

8. **Lower and middle school years allow you flexibility in the teaching approaches and assessment that you use.** This is because the results of assessment are used to rate students internally rather than externally. If a new type of assessment task doesn't work the first time, then change it and try again. You may well produce a great learning experience instead of an assessment task for your students.

9. **Share successes and disasters with your colleagues.** This process will become an informal professional development for you and your colleagues. You may, in fact, have an experienced colleague who can show you where you went wrong and how to overcome the error.

10. **Develop every skill you can in all your students, irrespective of their Maths talent.** The greater the range of skills you can teach your students, the greater is their chance of success in the long term.

11. **Help students develop their own understanding of Maths, not adopt yours.** In other words, introduce the ideal of 'Constructivism' into your teaching.

12. **Model aloud how you actually think about a problem/exercise.** Don't be the 'perfect' Mathematician. Include in your modeling any ideas that come to mind that you reject. Explain why you rejected those ideas. Model as many different solutions or approaches as time permits. If a student comes up with a different but mathematically correct solution, then have them convey it to the class.
13. **Challenge yourself to help students want to come to Maths lessons.** On the other hand, create a personal mindset that helps you develop lessons that you enjoy providing for your students. This means that you would want to be there.

Late in my career, I became a Head of a Mathematics Department in a large school. I found that many of my teachers were bored teaching Mathematics. During the next fifteen years, we had to introduce several new syllabuses with new teaching pedagogies and assessment techniques mandated. That forced us to look at what we did in the class room.

The ideas suggested above came out of that review. What occurred as a result of this process was twofold. First, our students became more committed to their Maths and the behaviour problems reduced dramatically. Secondly, teachers began to enjoy their teaching again.

USING THE TEXTBOOK AS A TEACHING/LEARNING TOOL

At the outset, let me say that teaching from a textbook is a 'no-no'. It would almost always give the impression to students that you don't know your 'stuff' or you are not confident. Here, you need to remember that you will always know or understand more than the vast majority of your students with the topic you are teaching. There will occasionally be gifted students in your class who will understand it all. Use them as tutors/mentors for fellow students. They will often express difficult ideas in ways better understood by their classmates.

However, since textbooks can be expensive, it is important that they be used often and effectively. A good textbook can be a real asset to you, as well as for the student. The author was, most likely, a teacher who has written the book on the basis of their own experience.

A good text is an essential tool in learning, helping with consolidation and practice of skills. Therefore, take a great deal of time researching various choices to find the one that best suits your needs. A bad choice becomes an expensive mistake.

The writer spent most of his career teaching Mathematics where a textbook was essential. During his years as Head of Mathematics, he introduced several new syllabuses into his school. Consequently, he and his staff needed to research carefully to find the best available textbooks. Sometimes it took up to two years to test out the available textbooks. His experience in using textbooks and selecting new ones provided the information for this chapter.

A good textbook is a:

- # Reference guide;
- # Source of the basics to learn and/or practise;
- # Source of rules or procedures to be learnt;
- # Source of exercises used to consolidate new learning and ideas on how to practice new skills; (These need to be carefully graded to allow students enough practice to consolidate the basics and then allow them easy movement forward to the more challenging tasks.)
- # Source of revision exercises;
- # Source of worked examples; and
- # As a backup when you teach new and challenging topics.

It should contain the following:

- # Chapter summary of ideas to be remembered;
- # Chapter review/test;
- # Answers;
- # Easy to understand diagrams, graphs, pictures and other illustrations; and
- # It may also contain a suggested work program as a guide to the time you need to spend on each topic.

Teach your students how to use the textbook effectively as it can be an efficient learning tool at home. Spend a lesson looking at the following list of sections in the textbook, explaining how to use each part.

(a) Contents page

(b) Index

(c) Explanations of new work

(d) Worked examples

(e) Learning work

(f) Exercises to do

(g) Skill practice

(h) Chapter review

(i) Chapter test/diagnosis

(j) Problem solving

(k) Extension work

(l) BOB, back of book — answers

Work Program and Your Textbook

Give students a work outline to go with the textbook. In it, detail the basic exercises the students need to do to gain an understanding of the subject. This is the minimum requirement only. However, specify what needs to be done to extend this understanding to gain the best marks possible.

Some final advice:

Insist that students have their textbook with them every day in class. If you know that you will definitely not be using it next lesson, tell the class not to bring it. (Students appreciate not having to carry heavy books unnecessarily.) However, don't forget to remind the class to bring it to the next lesson. You need to be careful with this process in lower level classes, especially in junior high school classes.

When you know that you will use a textbook in a lesson, insist it remains closed until you have finished the teaching part of the lesson unless you plan to refer to it as part of your teaching strategy. It is important to write the number/s of the page/s you intend to use on the board. Then, don't start referring to what the students must do until you are sure everyone is on the same page. This is also true when you set work from a textbook for the students to do.

Some teachers, particularly, in lower high school classes with less able students, have extra textbooks with them or photocopies of the relevant pages to ensure all students will be able to do the planned work.

In the end, your text must be:

- A teaching tool;
- A teaching aid; and
- A learning tool

AN APPROACH TO PROBLEM SOLVING IN MATHEMATICS

The key to teaching Problem Solving in unfamiliar contexts is to begin as early as possible with ***all*** students. Give your students as much experience as possible as often as possible. This experience does not need to be for extended periods but should contain many short, sharp exercises in a great variety of contexts.

What we need to do is to build up our students' confidence by giving them lots of opportunities, rewarding them, ***not*** so much for a correct answer but for being involved in ***'having a go'***.

Here are themes/ideas that I use to develop this confidence:

- Adopt the premise that problems are 'easy'. Teach your students to start by looking for a simple approach.
- The word 'problem' has a negative connotation. Perhaps the word 'challenges' is a better way to speak about our 'problems' in unfamiliar situations.
- Mistakes are to be welcomed. They are learning experiences.
- Finding a dead end in an exercise should be regarded as a success not a failure. You have just proved you cannot do it that way.

The seeds of the solution are always in the problem but often students overlook the obvious. So teach students how to interpret questions.

I would often model verbally and on the board how I approach a problem and work towards a solution.

I encourage my students to find and share different ways to solve the problems.

Persistence is an important habit to develop. Include in your work, occasionally, long problems to solve. They don't need to be difficult but just have many steps.

We need to give students greater opportunities to follow through with a problem. Give hints rather than solutions initially. Solutions should be given when all else has failed. Allow successful students to explain the solution to the class or group.

Charles Lovitt, a well-known Australian researcher into Mathematics teaching, has a thesis that goes like this: **'It's not the question that is important *but* how you ask the question.'** Mental Arithmetic in Middle and Junior High School is a great way to ask the same question in a variety of ways. I use it often.

Here are some simple examples of the one question to explain what Charles Lovitt meant:

7 plus 5

Find the sum of 7 and 5

What do I add to 7 to get 12?

What do I subtract from 12 to get 7?

What numbers have a sum of 12?

What numbers, whose difference is 2, have a sum of 12?

One last point

I do all the problem-solving 'challenges' I give first, before I give it to my class or I treat it as an unseen problem and do it for my class showing them that I am human and can be 'stumped' initially. Then, I treat it as a class exercise.

USING THE QUIZ AS A TEACHING STRATEGY

In recent years, there has been much discussion in the outside community that students are leaving school with less knowledge than their parents. There have been calls to return to the old ways of teaching which would include the use of a class quiz from time to time. However, what parents and others have forgotten is that, right from primary school, students are now exposed to a greater number of learning areas than the old 'Three Rs'.

Consequently, the students have less time on the traditional subjects than was the case in years gone by. Thus, we teachers still need to look for efficient ways to ensure students get all the necessary essential learning so that they are ready to enter the real world at school's end.

Syllabus writers have tried to counter this loss of subject time by reducing content and therefore encouraging teachers to concentrate on teaching for understanding rather than for a body of facts. However, there is still, as ever, a need for students to know basic facts on which to build their learning and to have an understanding of the language of the subject area they study. The quiz is an effective way to enhance this learning and understanding.

The strengths of a quiz are many

1. They are time efficient. They can be as little as 5 minutes long or as long as 15 minutes.
2. The teacher can create a quiz for each topic, of questions that are often done poorly by students.
3. These quizzes can be recorded and used over and over again.
4. The teachers can ask students as a homework exercise to write quiz questions on the topic which they find difficult. These questions then become a revision exercise plus extra quiz questions that the teachers can use in the future. These questions also lead the teacher towards areas of the topic that the students find more difficult.
5. The quiz also allows young teachers a great opportunity to develop and practice their questioning skills.

The quiz can help develop:

better basic skills and subject content knowledge;
better problem-solving skills;
an understanding of subject terminology;
a fun environment;
listening skills;
a desire for more success;
self-discipline;
learning through the back door;
better self-esteem;
learning through competition; and
more self-confidence.

A quiz can be used:

- # as a revision lesson (part or all of one);
- # as a review of a topic before the teacher begins a new lesson;
- # as a review of a lesson just taught;
- # as a development lesson into a new topic, that is, questions are used to develop new ideas in a subject area by using students' own personal experiences;
- # as a diagnostic test;
- # to develop a better understanding of a subject's language;
- # to develop better communication skills in subjects;
- # to allow the teacher to find areas of learning that need to be revisited.

USING BINGO TO DEVELOP NUMBER FACTS IN MIDDLE SCHOOL MATHS

Bingo is a game I have adapted to use in Middle School Mathematics to revise number facts and other mathematical ideas. It is a fun way of getting students involved in improving their Mathematics.

You will need to use cards from a commercially produced *Bingo* game designed for class work or create your own cards on 3 X 3 grid. Nine numbers are enough otherwise the game may take too much time. Below is the way that I used *Bingo*.

Step 1:

All students number 1–100 (or a number you decide) down the left side of a page in their Maths pad. (A second column may be needed).

Step 2:

Each student gets a *Bingo* card and draws the card on the opposite page to the numbers. All cards are returned to the teacher before the game begins.

Step 3:

Instructions to students when the question is given:

(a) Work out every answer (mentally/on paper).

(b) Write the question against the answer in the list.

(c) Check your 'card'.

(d) Cross numbers on your 'card' equal to the answer.

(e) Raise your hand when you have a full line—vertical, horizontal or diagonal—or a more complicated 'answer', for example, corners.

Step 4:

The teacher obtains a number using a random number generator on a scientific calculator. Then the teacher asks a question/number fact from the list and records the answer. (Alternatively, teachers could make up and record new questions as they go along. I often did this so I obtained, over time, sets of questions I could use).

Step 5:

When students claim they have a line or some other 'answer' ask them to read out the numbers to check or check personally. If they are correct, they win. If they are not correct, continue until a student is correct. Don't acknowledge the right or wrong answer as some students may mark numbers they don't have in an effort to win. The game can end here. You could review all the questions especially any trick questions.

If the teacher uses a limited number of cards, then more than one student could have a correct card. (I often only used multiple copies of four different cards. However, I made sure no students with the same card were seated side by side).

Step 6:

Continue the process until someone has a full card and check again.

An optional procedure

Often you will find you can go through all your questions and no one will get a full card especially if you have some 'trick' questions. Try to discover which numbers are still uncrossed. These are either the 'trick' questions or a weakness in your class. Re-teach where necessary.

Note: Students may need to 'protect' their answers to prevent 'cheating', even though each child also has a different card. Then the randomness of the numbers selected and a student's skill level will decide the winner.

Other ideas you might like to explore:

- Use the 4 operations.
- Use Maths vocabulary that implies a number, e.g. decade, centimetre.
- Use the numbers 1 to 99.
- A highlighter is useful for students to use.
- Give the students their own blank 3x3 grid. Let them select their own numbers.
- Have the students who have only one number to go to 'win' stand up. This will add excitement.
- If you create single laminated cards, the students can reuse the same card for the next game.
    - The teacher can gradually increase the difficulty of the game by introducing different winning scenarios, for example, any line; top, middle or bottom line; corners; diagonals and, of course, the whole card. The teacher will need to demonstrate on the board using a grid what is required before each game is begun.

The game of *Bingo* has several advantages for the teacher. It helps:

Improve basic skills such as number facts;

Develop mental discipline;

Develop listening skills;

Develop perseverance/persistence

Add fun and participation to Maths.

USING BINGO TO CONSOLIDATE BASIC MATHS SKILLS IN LOWER YEAR LEVELS

Bingo is a game you can modify to use in your class to improve basic facts in Mathematics in a fun, non-threatening way. I call it, 'creating learning by stealth'. The students don't realise that they are actually learning.

Here are some ideas on how to proceed:

1. Start by just using numbers that come up randomly (see *Point 10* below). You need to say the number slowly and repeat it.
2. Your next step is to use simple Maths problems, for example. one less than or one more than or the next number, as starting points. Don't forget to speak slowly and repeat the question two or three times.
3. Try a 3x3 grid only.
4. Have multiple copies of about four different *Bingo* cards you have created on an A4 sheet.
5. Give each student one sheet and indicate which grid each child uses for each game.

6. Teach the class how to listen actively and how to concentrate. Revisit these skills each time you have an oral game or quiz.
7. Use 'eyes down' as a cue to listen intently. Tell the class 'eyes down' also means 'Don't talk or look around at others in the class.'
8. Demonstrate on the board what you mean by a line in *Bingo*.
9. Develop your own sets of questions that fit into your work program.
10. Buy a simple scientific calculator that has a random number generator and use it to create your numbers. Make sure you record them as you go.
11. You'll need to have a page with the numbers you intend to use so you can mark them off as you go ready for checking when a child calls '*Bingo*'. Laminate it so you can reuse it.

Outcomes:

For me, there are several positive outcomes. They are:

1. Creating better basic skills;
2. Revising number facts in fun way;
3. Improving mental discipline;
4. Improving listening skills; and
5. Creating success for less able students in a non-threatening 'evaluation' exercise.

Other ideas you might like to explore are detailed late in the previous chapter. These ideas can stimulate learning and add more excitement.

THE STUDENT QUIZ

Early in my teaching career, I was always looking for something different to stimulate student learning.

I found the simple quiz was a great diversion for the students from the normal chalk and talk lesson of that era. Therefore, I created a series of different quizzes that I used in a variety of subjects that I taught in lower high school classes.

This is one of those quizzes. It is called the *Student's Quiz* and I have included the two versions I have used.

Essentially, the individual student or groups of students develop the questions and become the quizmaster.

Here is the procedure for the two versions.

Version One

1. Select a topic. This may be one you have just taught or it could be one which needs revision.
2. As homework, the students are instructed to devise five questions each (with answers). These are to be written out neatly.
3. During the next lesson, the teacher asks the students for their questions, checks them and the answers and, if they are satisfactory, the teacher asks each student to give the class the test.
4. Ensure that the student delivers the questions in a way that the class can hear and understand. Check volume and speed of delivery and that there is enough time between each question to allow the students to write their answers.
5. The student gives and explains the answer.
6. The teacher adds any teaching comments and/or supports and/or corrects the answer given by the student.
7. The process is repeated as often as the teacher wants in the time available.
8. The teacher must check the questions of the next student to ensure that no question is repeated and that all questions are suitable.
9. To ensure that every student get a chance to ask a question during the one lesson, I sometimes allowed each student to select only one of their questions to give to the class.

Version Two

1. Divide your class into groups of four or five.
2. Each group is given a different topic, for example, a selected topic for the forthcoming exam.
3. Each member of the group devises five questions of varying difficulty on the topic as a homework exercise. Answers must be included.
4. In class the next day, each group test the questions on each other and then develop a five-question quiz on their topic—the questions should range from easy to difficult, with answers included. (All group members get a copy of their group's quiz.)
5. The teacher rearranges the whole class into the same number of groups but this time each new group consists of one person with questions from each of the previous groups.
6. Each member of the new group 'quizzes' their new group with their questions. This process may take more than one lesson but would allow the revision of several topics.
7. The teacher needs to roam the classroom, keeping the students on task and clearing up any problems.
8. The teacher is given a copy of each group's questions.

Outcomes that can occur:

Better understanding is created by:

1. individual question creation
   (a) group discussion of the best questions and the correct answers.
   (b) Students have ownership of the questions.
2. Several topics can be covered.
3. Students get experience in oral work and in explaining their Maths.

4. Students gain more confidence in their various subjects.
5. Students want to have the 'best' or 'trickiest' questions. This is a great motivation for many students.
6. The group acts as the 'correction mechanism' for errors in question technique, understanding of the students' learning and the answers.
7. Students enjoy their learning.
8. The teacher may gain a valuable reservoir of questions and answers.
9. The questions asked are often ones which students feel they need or want to know—almost a self-diagnosis.

A final observation:

Students invariably ask questions which are harder than those of the teacher so it is important to instruct the students to write five questions which vary from ***easy*** to ***difficult.***

THE FUN PROBLEM-SOLVING QUIZ USING A MENTAL ARITHMETIC APPROACH

You could use this process for any subject. Obviously, Maths is a key area for quizzes. I have included some Maths questions below as well as some non-Maths ones to give you some idea of how I use 'The Quiz'. You, too, can build up your own bank of questions to use.

Remember, questions need to range from the simple to the complex or more difficult. For difficult questions, I will often give clues and/or more points for a correct answer. The mental arithmetic approach taken in the quiz is designed to encourage students to become powerful listeners and to give them experience in problem solving/critical thinking in a fun, non-threatening environment.

The Procedure

1. The quiz can be a competition between boys and girls.
2. Points are given for correct answers and totaled for boys and girls after each question.
3. Points can be doubled or tripled for more difficult questions.
4. Students who are caught cheating have double points deducted from their team score.

5. Anyone with a pencil in hand when the answer is given is declared wrong. This is to prevent students writing in the answer and claiming a right answer. No ticks are required for correct answers or crosses for an incorrect one.

6. Use the mental arithmetic procedure to create a discipline environment and to encourage students to develop their listening and watching skills. (See below for the mental arithmetic procedure).

7. For a critical thinking/problem solving quiz, give the answer after each question and explain how you arrived at the answer.

8. Try to make these sorts of questions fun as well as challenging. Questions with a twist and which require concentrated listening work best. The questions don't need 'difficult' thinking all the time but rather require different thinking.

9. Below are two fun Maths questions to give you an idea of what I mean.

   (a) One day I saw a horse standing on three legs so I weighed him. I found he weighed 300 kilograms. The next day he was standing on four legs so again I put him on the scales. What do you think he weighed on four legs? (Most answer 400 kg. The answer is 300 kg.)

   (b) How many months of the year have 30 days? (Most say 4. The answer is 11 as February is the only month that doesn't have 30 days.)

10. Here are two non-Maths questions.

   (a) If you spell out the numbers (one by one from 1) how far would you have to go before you found the letter A? (Answer: 1000)

   (b) You have one match. You enter a room that contains a wood-burning stove, a kerosene lamp, and a fire place. Which do you light first? (Answer: The match).

The Mental Arithmetic Procedure

There are the rules that students are given when doing the mental arithmetic quiz.

First, the students:

(a) Write the numbers 1–10 vertically, one under the other, on your pads. (10 can be any number you have time for);

(b) Put your pencils on your pad; and

(c) Find something to put over your answers to hide them from roaming eyes.

Before the questions are given, students are given these instructions:

(a) Arms are folded—questions are given.

(b) When a student knows the answer, that student puts their hands on their head.

(c) When enough students have the answer, the teacher calls, 'Answers down.'

(d) When the answers are written, teacher calls, 'Pencils down.' The student then covers the answer, puts pencil down and again folds arms.

(e) The answer is given and the score for each team is noted on the board.

(f) When all the questions are completed, the answers are corrected and results totaled; the teacher multiplies the boys' score by the number of girls to get the boys' final total and does the reverse to get the girls' final score. Then the winning team is announced.

Please Note:

1. The 'arms folded, hands on head' instructions are designed to develop self-discipline as well as mental discipline. (They prevent the students picking up their pencils to work it out).
2. The time you give students for each question depends on the difficulty as well as the students' ability.

3. Wherever possible, teach students shortcuts and/or ways to check/estimate answers.
4. Teach your students listening skills, for example, look directly at the teacher, watching the teacher's face/mouth and body language.
5. Don't repeat questions more than twice unless a mistake has been made.

I have found that students in primary school and lower secondary school classes love the competitive nature of *The Mental Arithmetic Quiz* as well as the different types of questions from the normal ones they are given. You can add a further dimension to the quiz by challenging the students to aim at getting 100% correct. When you do this, begin with easy questions first and gradually increase the difficulty as you go. This will keep more students involved in this challenge. After each question, ask who still has 100% correct and continue to challenge those with one wrong to aim to get no more wrong.

I have used a short quiz to start the lesson or one to finish the lesson when I have time left. How long or short the quiz is depends on the time you have available or the purpose of the quiz. My advice is to have it at random times and as a reward for the class.

DIFFERENT QUESTION–SAME ANSWER

One of the skills of excellent teaching is good questioning technique. I was fortunate to be trained initially as a primary school teacher where you needed to be able to ask relevant questions in a range of subject areas. A year after graduation, I was transferred into the secondary school arena where, for almost ten years, I taught a variety of subjects including English, Science, Mathematics, History and Geography with even a little Physical Education thrown in. (The remainder of my career was spent teaching Mathematics). This experience enhanced my understanding of the need to question in different ways.

Here are points to illustrate the need to be able to ask questions in different ways.

- # Questioning in different ways can extend students' understanding of the subject.
- # It can enhance their critical thinking/problem solving skills.
- # It can teach the vocabulary of the subject studied.
- # It can develop an understanding of and the use of the language and terminology of the subject.
- # It will help develop communication skills as well.

Here are simple questions to support these points.

1. 5 plus 7
2. What is the sum of 5 and 7?
3. Increase 5 by 7
4. What do I get when I add 5 to 7?
5. What is 5 more than 7?
6. Simplify 5 + 7
7. Find the missing number 7+ 5 = ...
8. Solve 7 + 5 = ...
9. Two items I want to buy are $5 and $7 each. How much money do I give the sales person?
10. Increase 5 by 7
11. If the answer to a sum of two numbers is 12; and one of the numbers is 7, what is the other number?
12. 7 take 5
13. 7 minus 5
14. Take 5 from 7
15. 7 subtract 5
16. Subtract 5 from 7
17. What is the different between 7 and 5?
18. Decrease 7 by 5
19. What must I increase 5 by to get 7?
20. What do I take/subtract from 7 to get 5?
21. I have $7 and I spend $5 on an ice cream. How much do I have left to spend on lollies?
22. 5 times 7

23. 5 multiplied by 7
24. Add 7 up 5 times
25. What is the product of 5 and 7?
26. The factors of the number I want are 5 and 7. What is the number?
27. Solve 5 x 7 = ?
28. Find the missing number: 5 x 7 = …
29. Simplify 5 x 7
30. 35 divided by 5
31. 35/5
32. If 5 is a factor of 35, what is the other factor?
33. What do I get when I divide 5 into 35?
34. How many times can I take 5 away from 35?
35. How many 5s do I add together to get 35?

Of course, in each *group* of questions the answer is the same. I'm sure in every topic area at each level teachers could devise a similar quiz. Obviously, it is an opportunity to revise and enlarge the vocabulary of Mathematics while enhancing learning.

Teachers should keep a record of these questions and enlarge their list as they go. Teachers could ask these questions in the form of a quiz in primary and lower secondary schools where they can also be valuable revision tools.

Same Problems — Different Level Responses

The same basic question used in gradually more complex contexts can require more sophisticated responses from students as they grow in years and experience. Charles Lovitt of the Curriculum Corporation in Australia discussed this idea at a conference of the Queensland Association of Mathematics Teachers. I put together a report on his idea for my staff and below are examples of what he said plus some ideas of my own.

His first example was called Eric the Sheep which I understand has been around for many years.

Eric the Sheep — A problem for Year 2 to Year 12

The basis of the problem is as follows:

A flock of sheep is lined up to be shorn. Eric is the last in the line and wishes to be shorn as early as possible. So each time the shearer grabs a sheep to be shorn he moves up two places in the line. How many other sheep are shorn before he gets 'done'?

Year 1 Students are given, say, 20+ counters to represent the sheep plus a toy sheep but first a role play takes place with six students plus one who is Eric. They act out the scenario:

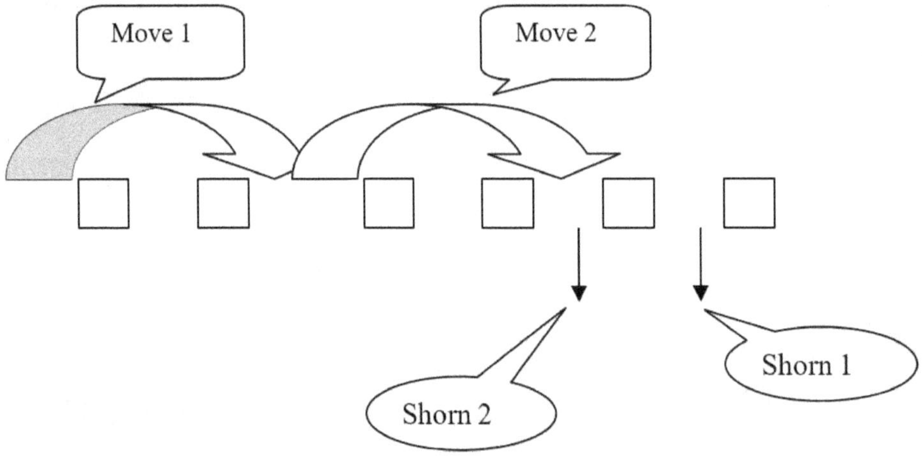

Year 4/5 Students are given 15 counters only and asked to determine how many sheep are shorn before Eric if there are 50 in front of him in the line.

Here the students must look for a pattern and hypothesise.

Year 7 The teacher changes the numbers 30, 50, 100, 1000 maybe even 'n' and asks them to develop a 'rule'.

Year 8/9 The problem is changed.

Next year Eric sneaks up 3 or 4 or 5 or 'x' places at a time with 'n' sheep. Develop a rule.

Year 12 The problem is developed.

The farmer has a good year so he can use 2 or 3 or 'y' shearers.
This gives us 3 variables (n, x and y).

Advantages of this Approach

1. It is open-ended.
2. It works from Simple to Complex
3. There are multiple exit points for different year levels.
4. It needs a model with little knowledge.
5. It is imaginative.

Some Measurement Problems

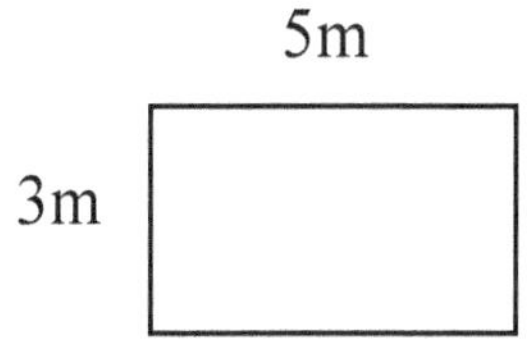

1. What is the area of this rectangle?
2. If I have a rectangle of area 24 square metres, what could the length and breadth be?
3. How many combinations are there? Are you sure? (Use whole numbers only)
4. Given 22 x one metre panels, find the largest rectangular area you can make with these panels.

ONE PROBLEM–MANY SOLUTIONS

It is important to teach students that not every question has only one correct solution, approach or answer. Of course, many do only have one. What is important here is that there may be more than one way to get to the answer/solution. In subjects like History and Geography, there may be a number of different interpretations of the questions which are equally valid. How students present facts in their discussions is what the teacher has to assess for correctness.

It is also important to understand that the quality of the answers your students give can vary within your class and in different year levels.

Below is a mathematical investigation that has been used in upper primary school to year 11 in secondary schools. The solutions vary from using counting with multilink cubes to using arithmetic progressions. I have included some of those solutions. Encourage your classes to look for and share different ways to solve or approach problems.

Skeleton Tower

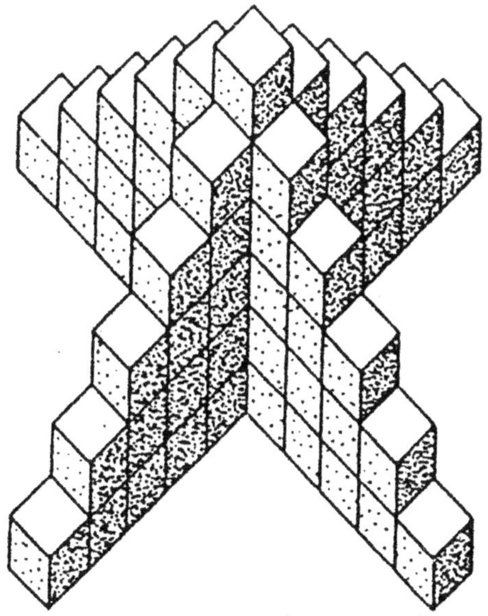

(a) How many cubes are needed to build this tower?

(b) How many cubes are needed to build a tower like this but 12 cubes high?

(c) Explain how you worked out your answer to part (b).

(d) How would you calculate the number of cubes needed for a tower ***n*** cubes high?

Some Solutions

At one school we found 11 different solutions to this investigation. Below I have included solutions which give you an idea of the range of the possible ways to gain a solution to most, if not all parts, except perhaps part (d).

The easy way to start is obviously to use the diagram or a three-dimensional model using multilink cubes or their like and just count. Certainly middle primary school students could solve the first three parts this way, as could the less able in higher school years.

The **first** solution requires the student to create a 3D model first and then break the model up as in the diagram. Then count the length and breadth of the rectangle and multiply to get the answer.

Solution **two** is similar to the first but with different rectangles. It uses a slightly more complicated arithmetic or, as you see in the diagram, a simple algebraic formula.

The **third** breaks the model up into four sets of steps and one rectangle (this being the central tower). Then a formula is created to gain the answer.

The **fourth** solution divides the model into horizontal levels. The students count each level and get the sum of all the levels. (The students may have worked out the relationship between each level).

In the **final** solution, the student uses an arithmetic progression to solve the investigation. This you might expect to be done in senior high school classes.

Solutions to Skeleton Towers are to be found in the Addenda (page 156).

MATHEMATICAL QUESTIONS WITHOUT NUMBERS

A Teaching Approach to Help Students Solve Problems in Mathematics

Students, as a whole, believe that a problem in Mathematics must have numbers before it can be solved. Students must come to understand that numbers are not important in deciding how a problem should be solved. They only become important after the student has decided how to solve the problem.

Let me give you an example. This question was put to a group of teachers at a professional development workshop. The teachers, initially, reacted the same way as the students did when I gave it to my classes. Here is the question, the reactions of my students and the teaching that took place afterwards.

The Swimming Pool Problem:

How long does it take to fill a swimming pool with a bucket?

The immediate reaction from most students and, initially, the teachers is to say, 'It can't be done. There is no data.' That is, you need the dimensions of the pool and bucket and how long it takes to fill the bucket. However, all of these are immaterial to actually solving the problem.

These are the steps in the solution that you need to do. They don't change because the dimensions change.

Work out the volume of the pool.

Work out the volume of the bucket.

Work out how long it takes to fill a bucket and pour it into the pool and return to the tap.

Work out how many buckets it will take to fill the pool.

Work out the total time by multiplying the time to fill a bucket by the number of buckets used. You would probably divide by 60 to find the number of hours to give a realistic answer.

Now you could use the dimensions to find the actual answer.

As part of the training of students to solve problems, teachers need to teach students to recognise the action words (verbs) that lead them to the operations they would use in solving the problem.

Here is one way to proceed.

Step 1:

To begin the process, select a series of normal written exercises in a Maths textbook. Ask your students to read each question and write down which of the four operations (addition, subtraction, multiplication, division) they would use to solve each question.

If the students were to use an operation more than once, they should indicate how many times they would use it. In lower grades, use only one operation at a time. Then, as you move up the school, increase the number of operations and the complexity of the questions used.

Step 2:

Select another set of exercises. This time remove the numbers and replace them with an empty box. Then, have the class repeat the process in step 1.

Step 3:

You then extend the complexity a bit further by insisting they put the operations in the order of their use. The questions are not solved. (This might be a good time to revise the concept of order convention and consolidate it in the students' minds.)

Step 4:

In upper primary and high school classes, use problems that would require the use of rules or formulae. Here they would simply write out the steps that they would need to take to solve the problem. This might mean the listing of formulae or rules they might use. *The Swimming Pool Problem* as used as an example earlier is the sort of question I mean.

What these steps are aiming to teach the students to do is to work out what are the steps they need to solve the problem.

When you begin a new topic with new exercises, have your students read the questions first and list the operations or steps they would use. Then, go through the early exercises with them to confirm what operations/steps they would take before you ask them to actually solve the problem and obtain an answer. This way, students will have the confidence to try to solve the new problems.

HINTS FOR THE TEACHING OF ALGEBRA

Algebra becomes, for some students in junior high school, 'the straw that broke the camel's back.' The students decide Algebra makes no sense to them and therefore Maths is no longer relevant either.

Over my years of teaching, I have developed a few 'tricks of the trade' that help alleviate, though not stop, this trend altogether. Remember that not all of us have a natural tendency towards thinking in a maths-logic way.

Here are my suggestions for you to try.

Algebra is just the next step in the number chain:

It is important to spend some time with your classes in junior high school showing the development of the number system from counting numbers to decimals to common fractions and so on right up to why we have algebraic numbers. As well, discuss the development of the four operations. Explain that the number system did not just occur. It really developed to keep pace with human development. Even the number zero had to be invented.

Algebraic numbers abide by the same rules as other numbers:

During this early time in junior high school, show your students how the rules for the four operations apply to all types numbers in the same way. For example, you can only add or subtract like terms. Certainly, these rules for the four operations are expanded slowly to allow the new number systems to fit in. Here, you can discuss working with power notation as an example.

The concept of an answer in Algebra:

This is the most confusing part for many students. 'Sir,' they say, 'the answer has no number value.' What you have to stress here is that they do have the answer. If they had a number value for the variable then they would have a specific answer. What you must get them to understand is that there can be an infinite number of answers depending on what the variable is.

Don't use just *x* and *y* as your variables:

Always use a variety of variables in the exercises you give your students to prevent the view that 'x' and 'y' are the only ones. Remember 'x' and 'y' are used in analytical geometry in your work in functions.

Making mistakes in algebra by students is more common than in other areas of Mathematics.

Mistakes in Algebra are made:

By failing to copy down the data correctly;

Not working down the page. (Doing that makes checking easier);

By using symbols incorrectly;

By poor setting out.

So make sure you model constantly the best way to set out your work to prevent unnecessary mistakes causing students loss of confidence in their ability to do Algebra.

Terms, expressions and equations

The very first lessons in formal Algebra will bring these three pieces of terminology to students. Spend time inculcating these ideas into your students' minds. Make sure you relate the four operations to them as well. I would do a lot of oral work using older traditional textbook material with them. A short quiz at the start of a lesson works well to reinforce the new ideas.

Equations:

These are the points that I stress constantly to my students when working with equations:

- # The aim of solving equations is to find the values of the variables that make the equations true.
- # The left-hand side of the equation is always equal to the right-hand side of the equation and vice versa.
- # In solving an equation, what you do to one side of the equation (in terms of the four operations) you must do to the other side of the equation to maintain the equality.
- # When you move one term from one side of the equation to the other side as part of the solving process, the operation sign associated with that term changes to the opposite sign, for example, plus become minus.

Finally, the last step in the solving process must be to test your solutions in the original equations.

Factorisation:

These are the points I stress.

Don't just use 'x' and 'y' as your variables.

In teaching new factorisation rules, show them in various forms, e.g. $ax + bx = (a + b)x$ or $xa + xb = x(a + b)$.

I stress that, when factoring, you are going from a sum or difference to a product.

It is important to treat expansions of products as importantly as the factorisation.

I stress that students should check their factorisation by reversing the process, that is, expand their answer to see if they get what they started with.

There is often more than one method to factorise an expression. So I encourage my students to find other ways and share them with the class.

Lastly, I stress that a simple idea, easily overlooked, can be the starting point to your factorisation process. Students often overlook the fact that $1 = 1 \times 1$ and $a = a \times 1 = 1 \times a$.

Products and power notation—expressions that cause confusion:

It is important to spend time ensuring students understand the difference between a product as opposed to a number in index or power form. I explain that a product is a way of expressing a number added up several times while a number in index form means the number is multiplied by itself a number of times.

Stress also the concept of *like terms* in relation to power notation.

Many students put Algebra in the 'too hard basket', even students who have been successful in the past. They give up quickly often because they can't see where the question is leading or what sort of answer is required.

The strategies above help give students a better start to Algebra so that they can gain the satisfaction of getting answers they didn't expect to get.

HINTS FOR THE TEACHING OF GEOMETRY

For your students to be successful in geometry, they **need to be taught to think logically.** Therefore, you should teach some basic logic. Here is an example of what I mean.

1. If $a = b$ and $b = c$ then $a = c$
2. If $a + b = c$ and $a + d = c$ then $b = d$
3. If $a = b$ then $a + c = b + c$; $a - c = b - c$; $a \times c = b \times c$; and $a / c = b / c$.

Secondly, **the language of geometry must be explained**. The derivation of the terms used in geometry often help explain their meaning. I often spend time discussing the Greek and Latin roots of the terms and then testing students' understanding of them. This will help in their interpretation of problems.

Now, once the students understand the terminology, you can begin on developing Euclidean Geometry from the initial assumptions and use them to prove the axioms you will use along the way.

Below are the ideas I feel are paramount to the teaching of Geometry.

Diagrams

Diagrams are the core to all your work in geometry. They are the means you need to develop your proof. Here is the advice I would give my students about how best to use a diagram. (Make sure you model all these ideas frequently).

The diagram should be large and to one side of the page so you can do the proof beside the diagram to assist your thinking.

Draw your original draft diagram in pencil and check it against the question as you re-read the question. If you are happy that all the data in your diagram is correct, then re-draw the diagram neatly in ink.

Now add any extra data to the diagram that is obvious from the data already available, for example, parallel lines can add lots of ideas. Some of this extra data may be unnecessary in the long run but it is important to note it anyway because it may, in fact, lead you to an alternative proof.

If you don't know where to start, then the place to start is always the diagram. Rotating the diagram can give you a different image of the problem to help you start.

Another idea is to look for the different shapes and position of angles to help you out.

Finally, you might use different colours to demonstrate ideas in the diagram that you might use.

Axioms

I instruct my students to list all the axioms they see in the diagram on the far side of the page away from the proof. Tell your students that all the information they need to provide for a proof will be in these axioms.

Working backwards

The student should ask themselves what do I have to prove last to complete my proof, for example, to prove lines are parallel, either alternate angles are equal; corresponding angles are equal or co-interior angles are supplementary.

If students do not know where to start, then teach them to work backwards. Teachers should demonstrate this often.

Different proofs

Geometry gives the teacher the opportunity to show students that there is often a variety of ways to solve a problem. Encourage your students to share different proofs. You might even like to create a contest where groups of students try to find the most proofs for a particular problem to reinforce this idea.

Constructions

Adding a simple line to a diagram can mean the difference between finding a proof or not. Thinking about what you might have to prove last to finish your proof will lead you to the type of construction you need to do that. Always pencil in the construction initially and use the diagram to go through your proof. Then, ink in your construction and add any axioms that come from the construction that assist your proof.

Assumptions

Euclidean Geometry is based on three assumptions. Explain how as you introduce geometry to your classes. Define exactly what is meant by an assumption.

The Development of Axioms

When you begin Euclidean Geometry, show how the assumptions and the terminology fit together to create/prove the first axiom, that is, the assumption, 'Adjacent angles forming a straight line are supplementary' is used to prove, 'Vertically opposite angles created by straight lines are equal in size.'

In lower high school classes, when geometry is the topic, I will always give a revision lesson where I develop each of the axioms (to be revised) beginning from the assumptions to remind students how they all fit together. It also shows that geometry is a continuum.

Model proofs

As with all parts of Mathematics, it is important for the teacher to regularly model proofs for students. Some teachers actually give students a 'model' to follow. It might be divided into these headings: *given; to prove; constructions*; and *proof*. Justification for each step would also be included as well as a detailed diagram.

Further hints for the student

\# Look for the obvious or the simple ideas/axioms to use to start. Don't assume you need a complicated idea to begin.

\# Work down the page. It is easier to check and reduces errors.

\# Remember that angles are named because of their (a) size; (b) shape and (c) position. Often they come in pairs.

\# Even an obvious fact must be justified.

Geometry is fun to teach. Often students find it difficult to see its relevance but the mere fact that it teaches logical thinking makes it an important learning experience for them.

COOPERATIVE LEARNING EXERCISE IN MATHEMATICS

The Five Blackboards Strategy

In Year Ten, I spend as much time as I can with the more talented classes teaching logical thinking and deductive reasoning in their Euclidean Geometry Unit.

Additionally, I spend time teaching my class a formal approach to geometric proof.

One of the activities I always do is to give my students a series of five written questions such as the one below on Circle Geometry.

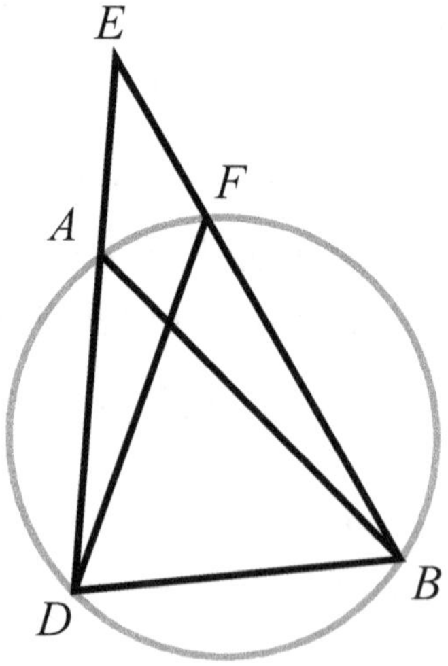

In ABD, DA is extended through A to a point E such that $AE = AB$. If the circle drawn through A, B and D cuts BE at F, prove that $FE = FD$.

This strategy has a number of steps and it is used after all the basic teaching and practice with circle geometry has been completed. It is the final step in my teaching of this unit.

Here is my strategy.

Step 1:

The students in the class are given 5 problems similar to the one above for homework. They must draw diagrams for each one and write a proof for each problem.

Step 2:

At the beginning of the next lesson (we have 70-minute lessons), I divide the class into groups of five, giving each student a number 1 to 5.

Step 3:

I have already organised five to six blackboard/whiteboard areas in my room or a nearby space. (I use mobile boards.) I allocate each group a board area.

Step 4:

Each student in a group has been given a number which is the number of the problem for which they will write a full proof with diagram on their board. They will explain their solution to their group as they go.

Step 5:

The students in each groups interact with the 'student'-teacher where necessary to gain more explanation or to debate the 'correctness' of the proof.

Step 6:

Once the 'student'-teachers completes their 'proofs', they rejoin their group to allow a new 'student'-teacher to give their proof to their problem.

Step 7:

The teacher moves from group to group getting involved only as a last resort.

Step 8:

Once the process is finished, the class gathers in their normal positions where the teacher asks the students to share any problems encountered or any innovative proofs.

Step 9:

The teacher reviews the process, highlighting any important issues not raised in Step 8. I feel my students always gained much from this process. Here are some of the comments they made during our evaluation of the activity.

1. I found I understood better because the student was explaining in 'our' language.
2. We all helped each other.
3. Others gave us good ideas.
4. It was more fun than a normal lesson.

For me, as the teacher, the positive outcomes were:

Communication skills were developing.

The students were on task willingly.

They had ownership of what they did.

Mentoring among students was beginning to develop.

This strategy could be used successfully in many topics; for example, trigonometry and simultaneous equations.

LESSON STARTERS

A good lesson is like a good speech. It has an opening that creates a desire on the part of the listener to wait to hear what the speaker has to say. Without this desire, the message of the speech will never take hold in the minds of the audience. The start of your lesson must create that desire in the minds of your students. Introductions to lessons must not be long, but must challenge the student to listen attentively and must create a desire to learn.

Below are a number of ways you might introduce your lessons to add interest to your topic and a desire on the students' behalf to learn all about that topic.

The Quiz: This can take any form you wish that fits into your teaching topic. It can be as simple as a quick revision of the last lesson you taught beginning with simple questions and ending with some tricky or difficult ones to interest all the class. The quiz should last only five to ten minutes, depending on the length of your lesson. Add interest with a competitive flavour. This works well in primary and middle school classes.

Joke, Cartoon or Riddle: Again, it must be related to your lesson topic. One riddle, cartoon or joke is enough. Try to insure the class don't go 'overboard' with 'canned' laughter as this will destroy the interest you hope to create.

A Problem-Solving/Critical Thinking Task: Based on the topic of your lesson, they should need only a short time to do and allow all students to get a start. You may need to give clues out slowly to help all students get started. Perhaps,

you may ask a student to explain their answer or solution to the class. Students speaking to a class create a greater interest in what you are doing. These students give the impression to the others in the class that they, too, might be able to succeed.

A Game: Teachers often create or use games that highlight ideas or skills to be learnt and understood, for example, the game of *Buzz* is a counting game used in lower primary schools to enhance the ability to count in a fun atmosphere. Games should be simple to play as well as fast and enjoyable.

Brain Gym: There are many sets of exercises within *Brain Gym* that are designed to awaken various parts of the brain that we use in different thinking processes in different learning situations. These sets of exercises take only about five minutes. I, personally, have used them with less able classes and have found they encourage longer concentration.

DVD, Video, CD, or Audiotape: A short piece from any of these could introduce a lesson to create interest. Be sure to cue it to begin at the correct starting place, your machine is in working order and is ready to go. Don't be tempted to continue beyond the point you were going to use, even if the students ask for more.

Students Talks: These can be an excellent start to a lesson. Students are almost always interested in what other students have to say, particularly in primary schools. Keep it a short activity. Make sure your student presenters are well prepared beforehand and give them as much encouragement as you can, as most students initially find it difficult to speak publicly. Teach your class to be a respectful audience with rules on how to behave as a member of an audience.

A Reading: This could be done by a student with good reading skills or by you. Obviously, it should be short, interesting and on your teaching topic. Practise beforehand, making sure you put expression and emotion in your voice to add further interest to the reading. A student reader should be given the passage early to practise beforehand.

A Short Story: Here are some ideas: in Maths, you could talk about the invention of new numbers when you introduce them first to your class, for example, negative numbers or common fractions. In History, there are many stories to be told. In Science, you could tell the story of a discovery, for example, gravity, the

apple and Newton. In English, you could discuss the origin of different words that you have introduced to your class. There are endless numbers of stories to be told. As you find a new one, add to your collection.

Homework Review: This could be simply a chance for students to ask questions on points they did not understand. Alternatively, it could be where you, as the teacher, consolidate the difficult or often misunderstood parts of the homework topic. Thirdly, you could simply have a homework quiz where you ask the class questions on the most frequently misunderstood ideas from that topic.

Last Lesson Review: This, in some ways, is a little like the homework review. It should contain a review of the main concepts as well as a discussion on the areas of the topic that students often don't understand or forget.

Question Session: As a homework exercise, you ask the class to write down questions they might ask about the last topic they studied. They should know the answer. The next day, each student asks you to answer the question. You could give a wrong answer to test them out. Alternatively, each student could ask the class a question whose answer they know. You may need to check the answer beforehand. The question must vary in difficulty. For you, the teacher, the questions are clues to what your class find difficult and therefore will lead you towards what needs to be reviewed or stressed as important. Finally, you will need to limit time on this approach.

Play On Words: These encourage students to think in different ways. Often, the so-called 'bright' students find these difficult while others latch onto the different thinking processes quickly. Again, five minutes is adequate if you have the *Play on Words* already written on the board. If you can find ones that have a connection to your topic, all the better.

Motivational Phrases or Sentences: I would often put one on the board before a lesson, especially in my high school classes. Sometimes, I would be asked to explain what it meant. Many students would write them in their diaries for the future. Here are some examples:

- # Goals that don't stretch you won't excite you.
- # You can't spell '**success**' without **U**.
- # Thinking is the hardest work of all—that's why so few of us do it!

Things to Remember

Limit the time to between five and ten minutes.

Relate your lesson starter to the theme/topic/context of the lesson you are about to teach.

Be well prepared so that the lesson starter gets the students involved quickly and easily.

Vary the ways you start lessons to add a further degree of expectation in the minds of your students.

If it works well, store it in a folio for that topic to use the next time it is appropriate.

Create a folio of lesson starters for future use.

Always use the 'KISS' principle—'Keep It Simple, Stupid'.

Be enthusiastic and show your love of the subject/s you teach.

MATHEMATICS TEACHING

—and—

TECHNOLOGY

A GUIDE TO USING TECHNOLOGY IN THE MATHS CLASSROOM

Using computers, software and calculators of all types

1. Always test your equipment/technology first to ensure it works.
2. Practise its use before the lesson.
3. Remember: You must still do the teaching. Technology does not teach. It simply assists the process of teaching/learning.
4. Make sure all students gain equal access to whatever technology you use.
5. Ensure that all students who need to use the technology can use it effectively and as efficiently as possible.
6. If you use technology as a medium in your assessment, make sure the success in the task is not dependent on the students' expertise in using the technology, as an expert.
7. Before you use any technology as part of your assessment task, make sure all students have practised and perfected the technological skills they need to succeed in the task, given that they know the appropriate learning work.

8. Develop an expertise in using all available technology in your classroom. This will give you opportunities to use a greater variety of teaching strategies/ pedagogies.
9. Always plan a back-up strategy if the technology fails, that is, have spare devices or an alternative teaching strategy.
10. Where you have competent students with technology skills, use them to assist you as demonstrators or as mentors for others.
11. When acquiring new technology, be sure to investigate the real cost including the 'behind the scenes' costs such as servicing, as well as the benefits to your students. Does it offer a better teaching result than the already available strategies/ technology?

Remember the advantages of technology: calculators (scientific and graphics) and computer software open Mathematics teaching and assessment to real life problems. This makes Mathematics more relevant to students and that is a great advance.

A GUIDE TO USING CALCULATORS IN THE CLASSROOM

The calculator, in its many forms, has made the life of the student less arduous than in the past. Yet it has its challenges for the teacher. Teachers must still teach the students the Mathematics while the students must understand the Mathematics of the procedure to gain the most benefit from this powerful tool.

The calculator takes the drudgery of long calculations away from students, allowing them to gain more experience of a topic in a much shorter time.

It also allows the teacher to bring more life-related exercises into the classroom, particularly in later years.

What is important for the teacher to remember is that the calculator will not do the teaching for you. It is merely a tool that you have your students use to help consolidate their learning.

Below are strategies that will help you, as a young teacher, to gain the maximum advantage from using calculators.

- **Have the students use the same calculator.** This makes it easier for you when you only have to give one set of instructions. This becomes more important especially with graphics calculators.
- **Have a large chart of the calculator face/dial to use to show where the keys you want to use are.** Leave the chart displayed on a noticeboard at the front of the room permanently to allow you to demonstrate when individual students need help.

\# **Use an overhead transparency with the calculator face on it to help in the same way.** This allows you to point to the keys you want the students to use while you face the class, giving you the opportunity to ensure all your students are on task and to see those who need your help.

\# **When you are teaching a new procedure, write the steps and symbols on the board.** Make sure every student does each step with you. Have each student check their partner's calculator as they go. This is especially true when using a graphics calculator.

\# **Teach your class to understand this axiom:**

The calculator is only as smart as its operator. *That means:* ***'rubbish in, rubbish out'.*** *Explain to the class, frequently, that the calculator gives perfect answers to whatever you give it so the student must press the correct buttons in the right order.*

\# **Make sure, as best you can, that the students understand the 'Maths' of the process they are using.** However, knowing the correct procedure to use will at least give them the correct answer.

\# **Encourage the class to do the calculation twice using the calculator to check that they get the same answer both times.** If they don't, they should redo it. Tell them to write the answers down each time so they can check them.

\# **Tell your class how to get an estimate of the answer as a way of checking that their calculator answer seems correct.** If the answer and the estimate are quite different, the student needs to investigate further.

\# Inform your class that **flat batteries can create abnormal results.**

\# **Use your best calculator students as mentors for those whose calculator skills are lacking.** Their language is the language of their peers who often get the 'message' more quickly from them than from you, the teacher.

\# **With graphics calculators, it is important to teach your students the 'reset' procedure** as, often, stored data will corrupt the calculations being done.

\# **For advice** on particular makes of graphics calculators and their special application, **the teacher need only go to the company's website** to get further support. These websites often have lesson plans to help the inexperienced teacher begin using these powerful learning tools.

Finally, **you must be very proficient in using the calculator.** This is especially true with the scientific and graphics calculators.

In conclusion, remember the calculator cannot do your teaching for you. In the end, the calculator, even with all its power, will need your input somewhere for it to be truly effective for your students.

A GUIDE TO CALCULATORS AND ASSESSMENT

Included below are points that a teacher should take note of in having students use calculators in assessment. These ideas come out of my experience of over 25 years of calculator use in the classroom plus the introduction of graphics calculators to my school during my years as Head of Department.

The first set of ideas is for calculators generally. The second set relates to graphics calculators.

Calculators generally

Calculators mean normally less time is required to complete exam tasks than in the past. Often, using calculators means less working needed to be shown.

Real life problems can be used especially with scientific and graphics calculators which have many different functions/procedures that students may use.

Calculator-wise students may find alternative ways/approaches to solve the task/problem. Be aware of this and test out the 'unexpected' approach/solution. (After the examination, demonstrate this approach/solution to your class. Better still, have the student do it for you.)

Students must be trained to 'estimate' what the answer might be before using their calculators (even with the four operations calculators). This should be an

automatic checking procedure that you, as the teacher, are trying to create as a habit in your students. Teach them to look at the answer they get to check that it is realistic and/or life-like.

Your assessment tasks should specify that calculators can/are to be used and students should be aware that they will need a calculator for the assessment task.

In preparation for assessment tasks that require calculator expertise, make sure the students are given practice in all the calculator skills/procedures that they may need to use in the assessment tasks. Remember that you are testing their Mathematics, first and foremost, and *not* their calculator skills.

You may, of course, consider calculator skills important enough to test them in a formal exam. In your subject it might be a prerequisite before other assessment tasks. It should not, as a general rule, have a significant bearing on the final student report.

Make sure you have extra calculators and batteries available for emergencies.

Graphics Calculators

Assessment using graphics calculators does *not* work effectively if:

(a) Calculators need to be shared. Different students take different times for the same task so the first user often gets extra time. The memory must be erased before each swap which takes time.

(b) Students have limited experience with them. Their task/s becomes mastering the calculator before they master the test instrument.

(c) Extra time has not been added to allow for calculator errors to be found and eliminated by the less able student.

(d) Students are not taught to understand the learning work first and foremost. Remember the calculator is only as 'bright/stupid' as the operator.

(e) The batteries are flat.

(f) Students or teachers don't know how to clear the memory.

(g) Students don't understand 'menus' etc.

(h) You, the teacher, don't do the test beforehand using the Graphics Calculator to test the time needed and the validity of the test.

(i) You decide the marking criteria and time before doing the test yourself.

(j) The task is too long for the allotted time.

Some Advice for the teacher

1. Start simply.
2. Give students a practice first on the task/scenario to be used. Use a different content.
3. Keep the task short, time-wise.
4. Allocate more time rather than less time to do the task.
5. Ensure that you have a practice activity beforehand that uses the calculator techniques needed for the assessment task.
6. Review and rewrite the task (if necessary) ASAP.

Use the reviewed/rewritten task next year or with a different class this year. Do not throw away this task because it didn't work the first time. Remember it was only your first attempt.

A GUIDE TO INTERNET AND ASSESSMENT

This section is designed as a starting point for teachers new to using the internet as a tool in their assessment. Being proficient at using the internet itself is not a guarantee you will immediately succeed in using it for assessment. Nor is being inexperienced with using the internet going to prevent you from using it to create an effective assessment task.

The advice that follows is the culmination of my experience introducing the internet as an assessment tool. It is important to understand that my staff and I began with varying degrees of experience with the internet.

Below is the advice I would offer to young teachers joining my staff in the first year of their career.

- Make sure the assessment technique is not a test of how to get to use the internet.
- Always do the test yourself first, looking for 'glitches'.
- Allow more time than you feel is necessary. Computers and networks abide by 'Murphy's Law'. Websites can often be unavailable due to an upgrading process.
- Have more than one internet site available for use, if possible.
- Always review the task after you have used it and rewrite it, if necessary. Don't disregard it because it didn't work well the first time. It may just need fine-tuning to make it a great assessment task. You may need to add extra websites or update your list.
- Try to write instruments that are procedural in nature and don't depend on the website to be visited. Thus, you can use the procedure over and over again changing only the topic/data and the websites to be used.

Remember to check your answers to the assessment task before marking as the website/s may have been upgraded since you wrote the assessment task.

Learn to use a variety of search tools, not just Google.

There are sites that offer assessment items that you might use or adapt, for example, **http://www.funbrain.com** has lots of quizzes on a variety of topics in Maths and other subjects that students could do on the internet. The site would email the teacher the results for the class. This site is still available as of 2026 and there are others of a similar structure.

It is important to check the websites you use for any bias. If you still wish to use a site, discuss the concept of bias with your class before you give the assessment task based on that website.

Where students are allowed to find their own websites, they must acknowledge them. This will allow you to test the integrity of the site, if necessary.

For lower year levels and the less able, make sure the websites you use or recommend are easy to use to facilitate the students getting started quickly on the assessment task itself.

It is important that the websites you use or recommend have reliable data. Check it yourself beforehand.

For the most reliable results, all students should have access to their own computer. Sharing computers can be more time-consuming and can lead to claims by students that they did not get equal time to do the assessment task. Inexperienced students do take longer to do the tasks.

Remember that you don't need to be an expert websurfer to get started. Start simply. If it doesn't work the first time, change it and try again.

You can always use the results as a guide rather than a final grade or not use them at all. It can simply become a learning experience, both for you and your class.

In conclusion, the internet offers you this great advantage. You get real life data and the relevance to life that your students so often crave in their learning situations.

GETTING STARTED WITH GRAPHICS CALCULATORS

Let me make this point first and make it strongly:

A teacher does not need to be an expert in using these calculators before it can be used as a teaching aid in a lesson. You need only the basic essentials to begin with for each necessary procedure.

Using an OHP unit with your calculator is a great way to begin. Then you have total control plus a class full of students just 'dying to help you out' if you press the wrong button. These calculators are becoming more user-friendly as time goes on. The reason is simple. The calculator manufacturers see educational institutions as a growing market. If teachers can't use them easily, then they will not use them.

Graphics Calculators, like computer software, are a very powerful tool for teachers. The advantage of this calculator over the computer is threefold—cost, portability and availability.

By *availability*, I mean that Mathematics teachers have to compete with every other subject department for computer resources whereas Graphics Calculators are mostly, but not only, in a Maths teacher's domain.

One of the great advantages of using these calculators is that you can develop an understanding of a topic, for example, Statistics, Graphing or Quadratics with little or none of the algebra and physical computations you need when using the chalk and talk/pen on paper approach. The teacher can develop the understanding first with the Graphics Calculator before the pen-on-paper process is started. This calculator enables you to do many examples quickly and visually thus using the visual and frequency needs of each learner.

Another advantage is that it is a great checking tool following a pen-on-paper example, for example, after a student has drawn a graph in Calculus using the 'old' pen-on-paper approach.

The Graphics Calculator makers provide excellent manuals and other publications that provide other teaching ideas which will expand your usage of these powerful machines. Each manufacturer has a website with further ideas.

Like all things newly learnt, start off with a small chunk and then use the well-proven learning techniques of frequency and recency to enhance the retention of what you have learnt.

Each time you try a new procedure with a class, practise first. Plan carefully and evaluate immediately after the event, noting errors and problems. Then do it again SOON. Keep refining your techniques and adding new ones to your repertoire.

One last important point to note is that students, especially boys, take to technology 'like ducks to water'. Therefore use their interest to create mentors for other students and yourself. Even use these students to demonstrate any new application you need to introduce into your program.

Remember the KISS principle—*Keep It Simple, Stupid*—initially. That way, you will gain confidence in using Graphics Calculators and they will become, for you, a valuable teaching aid and a great learning tool for your students.

GENERAL STRATEGIES

in

MATHEMATICS TEACHING

HOW TO HELP STUDENTS ELIMINATE MISTAKES IN MATHEMATICS

Confidence is a necessary part of being successful in school Mathematics.

However, there is a perception among many students that Maths is hard. Often, this perception comes out of the belief that, first of all, they can do their Maths only to find out that, in their assessment tasks, they get some wrong that they thought were right. More often than not, the wrong answers come from, not a lack of understanding, initially at least, but from a mistake they have made.

Therefore, as a teacher of Maths, I spend a lot of time showing students how to avoid many of the mistakes they make. When I teach a new unit, I explain where errors/mistakes are most likely to occur.

This chapter will list the generic errors and suggest ways to avoid them.

By teaching students where and how errors/mistakes occur, they will come to realise that Maths is not 'hard' and regain their confidence and achieve the success of which they are capable.

Below are listed the areas in which students make mistakes in their Mathematics with a short discussion on each.

Area 1: Copying down from the question

This is a mistake every student makes irrespective of their ability. It creates a situation where the question may not be able to be solved, thus wasting time and sapping confidence. The simple solution is to create within the student an automatic checking paradigm.

Area 2: Setting out

Poor setting out makes it difficult to find mistakes as well as follow through the process of your solution to find errors in your reasoning. Neatness, correct use of symbols and vertical setting out will help eliminate mistakes caused by poor setting out.

Area 3: Not answering the question/misreading the question

This, of course, is an obvious area in which to make a mistake. Teachers must instruct students how to read questions to determine what is required.

Area 4: Not answering every part of the question

This obviously follows on from Area 3 above. It occurs where a question has multiple parts. The student must be encouraged to cross off every part of every question as it is completed.

Area 5: Not addressing the assessment criteria

This is simply an extension of the ideas above. Not addressing the criteria not only creates poor results but wastes valuable time in examinations.

Area 6: Poor diagrams

Diagrams are supposed to aid the student in solving a problem, not confuse them. The diagrams must be neat, well-drawn and contain all the information from the problem. They must be large enough to see all the relevant information that will help solve the problem. In Geometry, in particular, I would have my class practise drawing diagrams for many problems without actually solving the problem. I would then draw a diagram for each to allow them to compare.

Area 7: Not testing solutions

When an answer has been found, the student needs to decide if the answer fits into what was expected. Students should consider the sort of answer they might expect when first reading the problem. In other words, the student should decide if the answer is realistic. Secondly, answers should always be tested in problems that involve solving equations. If students follow the correct procedure, then testing each solution in each equation will ensure they have the correct answer.

Area 8: Calculators

Calculators are only useful when used correctly. It is important that each time the calculator is used, the process is done a second time to check the first result.

The teacher's role in eliminating errors:

1. Reviewing bookwork

The teacher should create a model setting-out procedure for each type of Maths problem. Then, from time to time, it's necessary to review each student's workpad personally with the student to point out mistakes that are being made, why they are being made and how to avoid making them.

2. Assessment review

Here the teacher should draw the attention of students to mistakes made during an assessment task on the actual assessment paper. A personal review with the student is desirable, as well. In reviewing the assessment tasks with the whole class, the teacher should point out where all the mistakes are made for the benefit of all students. Lastly, as a motivational device, the teacher should indicate to each student how many marks they lost through avoidable mistakes and how those mistakes affected the final rating/grading of that assessment item.

3. Teaching to avoid mistakes

It is better to be safe than sorry. During the teaching of a new unit, I would always point out the pitfalls/places where mistakes occur. Sometimes in modelling answers, I would make mistakes deliberately to see if I could trick the students or to see if they recognised the mistake.

Obviously, there is some teaching to be done to avoid the areas mentioned above where mistakes are made, for example, teaching students how to read questions. The extra teaching you will need to do will only be found out by careful perusal of students' day-to-day work in class and at assessment time.

A GUIDE TO HOMEWORK IN HIGH SCHOOL MATHEMATICS

Most of what appears below was the advice that I wrote for teachers who taught Mathematics in my department when I was its Head. It appeared in my department's handbook.

Homework was an accepted part of what we did as Mathematics teachers for all classes except for those with special needs students.

How and When To Set Homework

- It should be set daily or after each lesson.
- Write the assigned homework on the board.
- Ensure the students write it in their school diaries at the end of the lesson. In junior classes, you may stand at the door checking the homework is written in their diary as they leave.
- Discuss how long the work should take and any necessary advice.
- Lastly, early in the school year, teach your students how to use their textbook to help them do their homework.

What Homework Should You Set?

For students to achieve their full potential in Mathematics at high school, homework must be done on a regular basis. Homework, based on current class work, is meant to be an extension of the lesson and is needed for the reinforcement of concepts.

In high schools, homework in Mathematics may consist of:

- Written exercises set for practice of skills and concepts. These are based on classwork.
- Learning work, for example, rules, vocabulary and theorems.
- Assessment tasks—these usually count towards Semester reports.

What about Students Who Don't Do Their Homework?

Teachers should record in their diaries the names of defaulters. Parents must be advised when a pattern of missing homework becomes evident. Teachers should develop a process for dealing with homework defaulters.

What If Students Can't Do Their Homework?

As most homework is based on the work done in class that day, this is not usually a problem for most students. However, if a student has difficulty in beginning homework, teach these strategies:

- The student should look for a similar problem in the work done in class. This is usually all that is needed to jog the memory.
- The student should look for an example in the textbook prior to the exercise. Each different type is usually done in full with an explanation.

- # If students still have difficulty, they should see their teacher the next day **before class** and arrange a time for individual help. Most teachers are available for 'homework help' at lunchtime or before and after school. Your teacher advises the student re availability.

What If Students Tell Parents That They Never Have Homework?

Often, there are complaints from parents who tell us that their students never have homework. This is clearly not the case! If a student has no written homework, (which is unlikely) then we would suggest that the parents set one of the following to be done:

- # Ask the student to write a summary of the rules for the current unit and to work an example of each type of problem. The textbook will be useful here. Look for chapter summaries.
- # Look at the student's exercise book and find an exercise that caused difficulty. Set this exercise to be done.
- # In each textbook, there are chapters on basic skills. Students can do any of the exercises from this chapter.
- # Often there are chapter reviews and practice tests. These can be done.

The Review Process

Homework should, wherever possible, be reviewed during the next lesson for the greatest impact on learning to occur. This learning may, in fact, be the basis of the next lesson. Even though there is a continuing debate as to the merits of homework, the advice here will help the Mathematics teacher deal with homework successfully.

REVIEWING HOMEWORK

Setting homework has no real value unless the teacher has a regular review process. Without this review, students will not see homework as important because they feel that the teacher doesn't see it as important enough to review.

Homework is designed to consolidate learning that has occurred in class. What your review does is to further consolidate that learning through the recency/frequency aspect of learning.

Below is how I went about reviewing homework.

First, there was my preparation.

I instructed my students to write down any questions that arose as they did their homework on the page where their homework was done, ready for them to ask during the next period's homework review.

Personally, I would review the homework tasks beforehand, deciding what I needed to emphasise or re-teach (if necessary). I might also prepare answers in hardcopy form, if I deemed it an advantage for my students.

Now, here is what I usually did at the start of the next lesson in reviewing homework.

I would check to see who had done their homework. (If students had not done or been able to do their homework, they were to see me before the lesson to discuss the issue).

I would look around at the rest of the class observing body language that might suggest that others had not done their homework.

If there were signs, then I would have all the students turn their books to the front so I could check. (This emphasised that I was serious about homework). I would note these defaulters and talk to them in their time.

Next, I would ask if there were any questions from the homework. These I would answer plus any follow-up questions.

I would model my answers to the class, that is, I would do this verbally and on the board. I might later provide a hard copy of my 'model' answer.

Following this, I would discuss any other issues I felt were important to stress about this topic, for example, misconceptions, frequent student errors and so on.

Next, after setting the work to be done by the class for this lesson, I would work with those students who say 'they don't understand'. Here, I would ask them where they 'got lost' and proceed, from that point, to re-teach the topic to them to create the best understanding I could for these students.

It is important to have a follow-up for the defaulters. A first step for them might be to do their homework during the next break and bring it to you at a specified time. Consistent defaulters might incur more serious consequences. I kept a diary record of all defaulters. This was useful at parent-teacher interviews and at reporting time.

Lastly, I never believed that students had no problems with their homework. I know from experience which areas would give students problems. Therefore, I would ask students questions in those areas as a review of the homework. I would highlight any other important points before I went on with the teaching program.

Remember:

The review of homework is as important as the students doing that homework, and

Homework is a component in the process of learning study skills. That is important for future academic success.

CREATING A REVISION PROGRAM

As teachers, we know that many of our students perform well in class but fail to show the effect of that learning when it comes to assessment. This is especially true where students do traditional pen-on-paper examinations that contribute towards exit certificates and university entrance. Therefore, teachers need to teach their students how to prepare for examinations so that they can show the examiner what they do know, not what they don't know.

One way you can do this is to create with your class a model revision program leading up to exam time and follow up and use the model with them.

Here is how I went about the process:

1. I set up my work program so that all the new teaching is completed at least two weeks before the examination.
2. During the last teaching period for new skills, I list the exam topics on the board and the lessons left prior to the exam that can be used for the revision program.
3. The class and I decide what topics will be revised and when and how much time we will spend on each topic.
4. The second last period of the revision program is set aside for a practice test.
5. The last period is used to review the practice test and answer any final questions.

6. During the second half of the lesson before the practice test, I spend time teaching or re-teaching what I call exam technique.
7. Once the program has been settled, I indicate where the rules, procedures and the learning work are to be found in the textbook, and that are essential preparation for the exam. As well, I indicate the basic questions to practise or answer in each topic. Finally, I explain what they need to do to get a better-than-average mark.
8. Then the students must revise each topic the night before it is set to be done in class, looking for and noting any problems they need to have addressed by me in the next period.
9. To begin each revision lesson, I might first have a short quiz of the basics the class needs to know. Then I answer questions from the class.
10. The remainder of the lesson is given to the students to continue the revision or to seek personal help from me.
11. Sometimes, a common problem arises so I re-teach it to the whole class or a group.
12. I insist that every student works alone on this revision unless I have asked someone to give help to another.
13. I never answer problems on different topics from the nominated one for that lesson.
14. I also tell the students they may see me outside class time for these questions and I indicate where to find me and when.
15. Where revision questions are available in the textbook, the students are referred to them rather than to the normal text questions to add 'newness' to doing the questions.
16. The practice test always contains questions from past papers/exams varying in standard from simple to difficult in the core work as well as in the extension/ problem-solving questions.
17. I always create a copy of the model answers which I publish on noticeboards or as a hard copy for each student.

18. I review the practice test, giving answers first and working or answering fully only the difficult ones or those requested by students.
19. I ask students to review the way they used their exam technique in the practice test and to work out how they can improve that aspect of the exam process.
20. In addition, I always point out where errors are usually made in all the questions so students are aware of what to watch for.
21. Finally, I emphasise that students should try every question in the practice test, doing the easy ones first to get a good start.

Using this process allows the teacher, especially the less experienced one, to get to understand where students have their greatest problems in each area that is taught. This gives the teacher the knowledge to address these issues with each new class in the years to come. In reality, it is a great form of professional development for any teacher.

PREPARING STUDENTS FOR HIGH SCHOOL MATHEMATICS

A High School Maths Teacher's Wish List

What has occurred in recent years as many more students complete high school and seek a tertiary education is a growth in parents wanting their children to do Mathematics at a higher level. They see Mathematics as a key to tertiary entry and insist that their children be given the opportunity to do the subject at most advanced level possible—even going against the school's advice on the matter.

Therefore, high school teachers of Mathematics must teach almost all students for all their years at high school, irrespective of their innate ability in the subject.

This trend will not go away and high school teachers need the help of primary teachers to prepare their students for the rigours of high school Mathematics.

This chapter is written based on my experience as both a high school Maths teacher and as a Head of Mathematics who often had to advise parents on what was best for their students in the subject. Much of what I write here was presented to primary school teachers in a workshop on the topic.

Most, if not all, of the points I make in this chapter will be known to experienced primary school teachers so it is aimed more at those new to the profession.

Mathematics is a subject discipline where students must develop their understanding of a broad range of topics. Learning rules and procedures can take the student only so far. It will not help in the modern world of real-life Maths problems in unfamiliar contexts.

To help prepare students for high school Maths, upper primary school teachers need to attempt to develop the following within their students.

1. A work ethic and one which is self-motivating. Often, students in Mathematics will need to work alone and unaided.
2. A homework ethic. The speed of teaching the syllabus requirements in high school is dictated by outside authorities. This means that the teacher must cover a mandated syllabus in a specific time. For students, this means that homework is an essential part of the learning process if they are to keep up with the pace of teaching.
3. A study ethic. It is important that students learn that homework does not equal study.
4. A belief that all students can do some Maths.
5. An understanding that Maths is an essential part of everyday life and we all do Mathematical things successfully every day, often automatically.
6. A belief in students that asking questions in Maths is a 'cool' thing to do.
7. A belief in students that Maths is not gender-specific, that is, not just for boys.

Below is a list of what I call essential preparation that is not directly Mathematical but will assist students greatly in their study of Mathematics as well as other subjects.

Students should be taught:

Study skills

How to be powerful listeners

How to ask questions

Checking procedures

Estimation as a checking device

Various problem-solving techniques

An effective setting-out procedure

That the answer only is not enough. The students must explain in written Mathematical form how they achieved their answer.

That there is often more than one way to solve a problem

An understanding of order convention

Examination technique

Communicating mathematically is a skill that needs to be taught. It involves students being taught the following:

1. The correct use of Mathematical terms including their spelling;
2. Correct use of all Mathematical symbols;
3. Logical setting out;
4. Justification of each step where necessary;
5. Logical reasoning;
6. The use of neat and clear figures, accurate and appropriate diagrams;
7. To work vertically down the page to allow ease of checking and the elimination of errors in copying;
8. The translation from one form of expression to another, for example. numerical/ verbal data to diagrams/tables/graphs/equations;
9. Correct and appropriate use of units, e.g. in area, volume and so on.

Lastly, you can give your students a taste of high school classes by doing the following. (You might call these suggestions an Action Plan).

Set your classroom up with desks in rows and teach a number of *Chalk and Talk* lessons.

Insist that students work on their own while doing Maths exercises in a quiet environment.

- # Use textbook exercises.
- # Run some formal, timed examinations in a formal classroom setting.
- # Do regular problem-solving exercises. Ones in unfamiliar contexts so they get accustomed to the idea that problem solving is an everyday event, not just one that comes up in assessment.

As I alluded to in the title of this chapter, this is a high school Maths teacher's *Wish List.* Whatever you can do as a primary teacher to help develop this wish list would be greatly appreciated by Maths teachers but, more importantly, will help students to step into the rigours of high school Maths more confidently.

THE STUDENT AND THE TRANSITION TO SENIOR MATHEMATICS

Every year, students will come to you as their Maths teacher in year 10 seeking advice on what level of Mathematics to select for their senior years at high school. It is a crucial decision because it can impact the choice of courses the student will have available to them if they seek to go to a tertiary institution.

Obviously, their talent for the subject is a consideration. However, many students in their junior years develop a dislike of Mathematics which leads to a lack of motivation in the subject. Many can't see its relevance to their future lives.

All is not lost for the future if a student selects not to do a Mathematics subject in the senior school or chooses a lesser version of the subject. Most tertiary institutions offer catch-up opportunities to those who find they want to, or need to, revive their education in Mathematics to proceed with the tertiary course of their choosing.

Initially, I would discuss with the individual student their past success, their strengths and weaknesses in the subject and what they need to do to remedy any deficiencies. We would also discuss their future career goals to see what level of Maths was appropriate for that career. Then I would make my suggestion.

Once I have given my advice to the student on which Maths subject to select, I would attempt to give them an understanding of what doing Senior Mathematics entails.

This is what I would say to individual students and to students as a whole in my role as Head of Mathematics.

It is your decision to remain at school and to choose the subjects you want to study.

Therefore you must accept ownership of your decision and the responsibility for your success or failure.

Year 11 brings to you:

A greater degree of freedom to choose when and what to do;

A greater expectation by teachers that you will accept the responsibility for your study and learning;

A chance to become an adult learner.

Year 11 also brings:

Distractions that are in the form of paid work and a social life outside school;

A steep rise in the level of understanding required to succeed in your study;

The need for constant hard work. Talent is not enough to guarantee your success. You will need to work harder and longer than you did in your earlier school years.

Year 11 can bring success to the student through:

- *Learning to work 'smart';*
- *Organising your time;*
- *Setting priorities, goals; (both short and long term);*
- *Communicating and working with your teachers; and*
- *Finding and using a mentor and being one for your fellow students.*

Remember these two ideas:

You decide to succeed or fail.

Your teacher can not do it for you. They are now only guides along the way.

Here are a few home truths I would like to leave you with:

What we have to learn, we learn by doing.

Time is like money. You can only spend it once.

Your purpose in life should be to do what you can as well as you can.

It is good to take an interest in the future. That's where you will spend the rest of your life.

Finally, whatever Maths subject they choose to do, they alone decide by their work ethic how successful they can be.

ADDENDA

MORE ABOUT THE AUTHOR

Rick Boyce began his career as a primary teacher where he learnt to teach across the curriculum.

He was transferred early in his career into high school where the next eight years he taught a variety of general subjects in lower high school classes. These subjects included Maths, Science, English, Geography and History. This gave him a wide experience in the various teaching pedagogies associated with these teaching areas. That meant that, when he was finally able to specialise in the teaching of Mathematics, he was not a traditional 'chalk and talk' Maths teacher. He used a variety of teaching approaches in his lessons.

This approach to the teaching of Mathematics stood him in good stead when he assumed the role of Head of Mathematics in a large school.

One of his responsibilities as Head of Maths was to inservice his staff as part of the introduction of several new syllabuses. To do this, he had first attend many professional development workshops. Being proactive in this regard saw him being appointed to Metropolitan North Regional Mathematics Advisory Committee to help with the introduction of these new junior high school syllabuses to all schools. He was also involved in a committee to create alternative assessment tasks for use with the new Year One to Ten Mathematics syllabus.

He was then appointed as teacher representative on the Queensland Board of Secondary Schools Studies Mathematics Advisory Committee. This committee was responsible for the writing and implementation of senior syllabuses. The work on this committee was a great professional development experience for him.

This experience and the soon-to-be implemented new senior syllabuses led him to see the need to organise workshops for teachers on how best to implement the new syllabuses. So at his school he organised these workshops for his staff and invited any interested schools to send their staff along as well. He made sure that he selected presenters that were well-spoken and had practical ideas that worked to give the workshops. Up to one hundred teachers could attend. These workshops began in 1992 and continued until his retirement in 2001. He, too, soon became a presenter with his special interest being developing problem-solving in Maths.

He also became a presenter at various other venues included regional school conference and the QAMT Annual Conferences.